SIMPLE LOW-COST WIRE ANTENNAS FOR RADIO AMATEURS

WILLIAM ORR, W6SAI and **STUART COWAN,** W2LX

RAC
RADIO AMATEUR CALLBOOK
P.O. BOX 2013 LAKEWOOD, NEW JERSEY 08701

Caution: Working on antennas or towers can be dangerous. All warnings on the equipment and all operating and use instructions should be adhered to. Make sure that the antenna is disconnected from the station equipment before you begin to work on it. Make sure that your antenna is not close to power lines, and that it cannot drop on a power line if wires or supports fail. Do not attempt to climb a tower without a safety belt. It is best to work on your antenna with someone who can assist you with tools and will be able to help in the event of a problem or an emergency.

Published in 1993 by Radio Amateur Callbook
(an imprint of Watson-Guptill Publications,
a division of BPI Communications, Inc.),
P.O. Box 2013, Lakewood, New Jersey 08701

Library of Congress Catalog Card Number: 76-190590
ISBN 0-8230-8707-7

Manufactured in the United States of America

5 6 7 8 9/01 00 99

TABLE OF CONTENTS

FOREWORD

Your antenna is a very important part of your amateur station. An efficient antenna means good signal reports, consistent QSO's and exciting DX contacts. The best transmitter is a useless toy if it is connected to a poor antenna.

This handbook is written for those radio amateurs who dream of a fifty acre "antenna farm" studded with beam antennas atop 90 foot telephone poles -- but who are frustrated by a thin pocketbook, fifty foot wide lots, unsympathetic neighbors, and steely-eyed building inspectors who hamper "sky-wire" ambitions!

Designed with an eye to ecomony and maximum performance, the antennas described in this book have been built, tested and used on the air by the author and his friends. These antennas are simple, inexpensive, and best of all, they work! Exact dimensions are given, and no guesswork is involved. Many of the antennas work efficiently on two or more amateur bands, thus giving bonus operation to the amateur restricted to a single antenna.

For a large chunk of money you can buy a good antenna -- but this handbook shows you how you can build your own efficient antennas at minimum cost -- and today this is important to all of us!

Metric system dimensions are added to all measurements in this Handbook as a convenience to the overseas amateurs. During the next decade the U.S. will be heading towards the Metric system and it is suggested the reader familiarize himself with the accurate and simple measurement technique. After all, wavelength is measured in meters, why not antenna dimensions?

The portions of this handbook dealing with basic antenna theory also appear in the handbook, *The Truth About CB Antennas,* also published by Radio Publications Inc. They provide a popular, non-technical introduction to the mysteries of antenna operation.

No matter how small your antenna farm, or how cramped your finances, this handbook can solve your antenna problem. Good luck and good DX to you!

Chapter 1

Sugar-Coated Antenna Fundamentals

(What You Don't Know Will Hurt You!)

Your antenna is your most important piece of radio equipment. It is the key to working DX and having solid, consistent contacts.

You have probably heard a lot of words and stories about transmitting antennas and some of them are true. "Sky wires" seem to be mysterious objects to many radio amateurs. This may be due, in part, to an incomplete understanding of how antennas work on the part of the perplexed operator, but it is also due to a lot of misinformation and "baloney" floating around in books, magazines and on the air.

It is true that the antenna makes a *big* difference in the performance of your station. If you are a skilled operator and happen upon excellent propagation conditions, you can probably work DX with a piece of wire lying on the floor of your shack, but you will do far better under ordinary propagation conditions with a good sky wire--so ignore the wiseacre who tells you, "Your antenna's unimportant; just string up a piece of wire!", or something along these lines. Such information is just as misleading as is the hard-sell advertisement of an aggressive antenna manufacturer who wants your dollars in exchange for his vastly overrated (and perhaps inferior!) beam antenna. Luckily for the amateur, such manufacturers are in the minority.

You will find, as have others, that a good working knowledge of antennas helps you to solve many of your operating problems, aids you in making a sensible choice if you buy an antenna, permits you to build your own antenna that will *work* and--in the long run--gives you a stronger signal.

This Handbook discusses practical antennas and gives you the basic information you need to have a top-notch antenna, regardless of the type you choose. Important dimensions for all antennas are given in feet and inches and also in meters and centimeters.

Simple and effective wire antennas for the HF and VHF amateur bands are covered in detail in language you can understand.

A brand-new chapter on "invisible" (low profile) antennas is provided for the amateur living in an apartment house or neighborhood that frowns on conventional, highly visible antenna systems. Mysterious matters such as standing wave ratio (SWR) and SWR meters are discussed and explained in simple terms. Finally, this Handbook provides complete, detailed data on an effective antenna tuner you can build.

In summary, then --

1- The antenna is a very important part of your station. A low power transmitter connected to a good antenna will outperform a high power transmitter hooked to a poor antenna.
2- A good antenna does not cost a bundle. You can build a good, efficient wire antenna for any amateur high frequency or VHF band for a few dollars.
3- A good wire antenna is *not* difficult to build or erect and it will launch your signal into space with a good boost. You do not need a lot of expensive real estate to erect a good antenna, either.
4- "Know-how" is *everything* in antennas (as in other things). Everything you want to know and need to know is in this Handbook in simple, clear and understandable language. No gobbledegook or double talk.

Let's start at the beginning. Antennas have had an interesting history over the years and a study of the past may prevent errors in the future. So here's a quick, concise background of the very early history of radio and a short introduction to radio waves, antennas and the nature of things. We'll look at some interesting early antennas, the sugar-coated theory of radio transmission, antenna gain, and so on. In addition, this painless introduction will help you to understand just how today's amateur DX antennas work. The straight dope, in other words!

The Early Days

Over 150 years ago experimenters found that when an electric current flowed through a wire, a magnetic field was found wrapped about the wire. Shortly thereafter, it was discovered that a *changing* magnetic field produced a flow of current in a nearby wire. From these two discoveries, Michael Faraday of England proposed the novel idea of a magnetic "flux field" or "lines of force" (invisible lines of tension in space: like stretched rubber bands) to explain the phenomenon of magnetic fields and force acting at a distance. Faraday, moreover, expanded his curious idea into a general "field theory" of force which proposed that all space was filled by various force fields: magnetic, electric, gravitational, and so on.

About 1850 a canny Scotsman, James Clerk Maxwell, derived a breath-taking concept of nature and revealed a striking set of mathematical rules that encompassed all known electromagnetic knowledge and in the broader sense predicted an entirely new theory of electromagnetic radiation, described by Maxwell in terms of his "field equations." Maxwell boldly stated that "light consists of undulations of the ether" and predicted that electric and magnetic phenomena were similar to light and that electric "undulations" could exist in free space, in the same manner as light waves. The electric "undulations" were described by a set of monumental equations that showed a wave freely travelling from place to place, with an interchange of energy constantly radiating outwards from the source.

This was quite an idea for 1850!

Hertzian Waves

Thirty-two years after Maxwell's amazing electromagnetic theory and forty years after Faraday's original suggestion of an electric field, Heinrich Hertz of Germany proved its existence. He built a powerful radio oscillator using a "sparking coil," Leyden jars and a simple antenna. The transmitter worked at a frequency of about 53 MHz (megahertz), just about where TV channel 2 is today! For a receiver, Hertz used a length of wire bent into a loop and having "sparking balls" at the gap (Figure 1). By painstaking adjustment of the gap, Hertz made his simple receiver sensitive enough to spark at a distance of about 30 feet from the oscillator. This amazing fellow then proceeded to focus his electromagnetic waves with simple directional (beam) antennas and reflect the waves from metal surfaces! Other experimenters duplicated Hertz's gear and soon extended the range of the sparking-ball receiver and sparking-coil transmitter up to three hundred feet or so.

An interested observer of these early experiments was Guglielmo Marconi, of Italy. In 1895 he started his famous experiments, culminating in his historic trans-Atlantic radio transmissions in 1901. "Wireless" had come of age and by World War I, Marconi radio equipment was placed in service by various countries and used to handle messages over hundreds of miles. Contrary to the work of Hertz, early

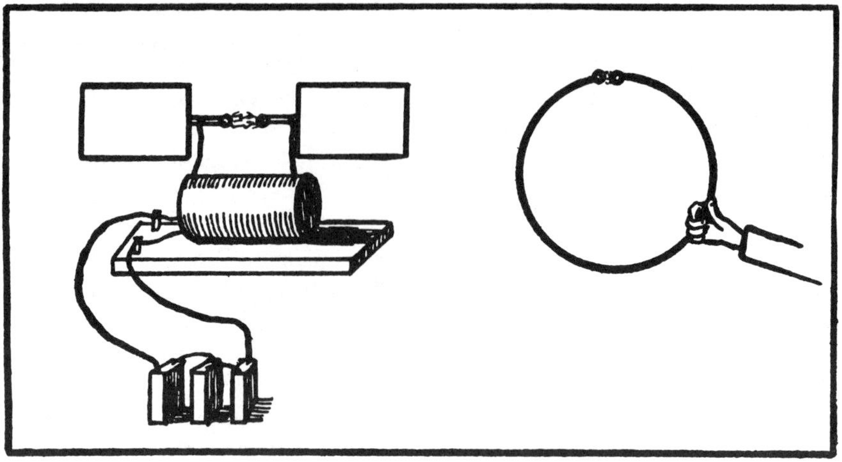

Fig. 1 **WORLD'S FIRST** radio transmitter (left) and radio receiver (right). In 1884, Heinrich Hertz of Germany generated and detected radio waves using this equipment. The waves came from a spark transmitter which used two copper plates as an antenna. Holding a resonant length of wire bent into a loop, Hertz moved about his laboratory and found that a small spark would jump across the gap in the loop within certain distances from the transmitter. In this fashion Hertz verified the formulas of Maxwell and proved radio waves existed and determined their wavelength. He also found that he could reflect and refract his waves with large metal sheets serving as a "radio mirror" (Drawing adapted from "Radio Theory and Operating", by Loomis, 1925).

"wireless" stations used very long waves, since it was obvious to these early experimenters that long waves were needed to cover long distances!

The Great Days

After World War I, the first radio amateurs experimented with "short" radio waves and frequencies as high as today's TV channels were tried by 1925. A great expansion of communication activity into the shortwave (high frequency) radio spectrum occurred when it was found that the waves could be sent around the world by bouncing them off an ionized layer of the atmosphere and back to earth. This layer (the *ionosphere*) varies in height 100 to 250 miles above the earth. With the perfection of stable frequency control for transmitters and

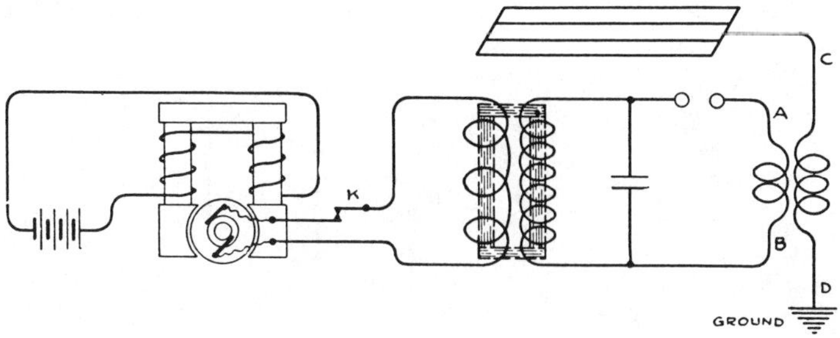

Fig. 2 SPARK RADIO TRANSMITTER of 1910 was run from a large wet battery and motor generator (left). A step-up transformer (center) provided high voltage for the spark gap and oscillation transformer (A-B). A huge cage antenna provided a communication range of 50 miles or so, using Morse code. (Drawing adapted from "Robison's Manual of Radio Telegraphy", 1918).

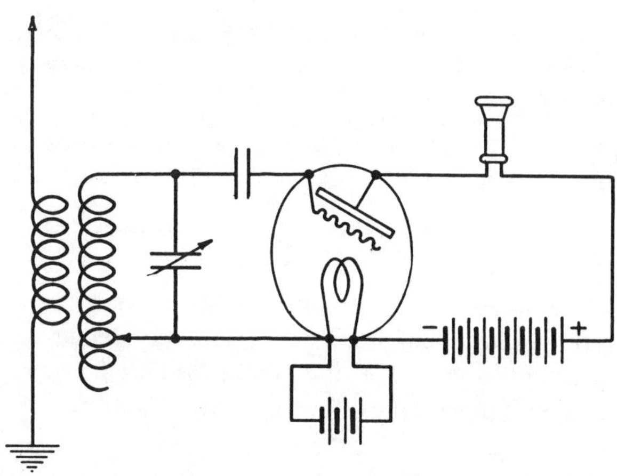

Fig. 3 FIRST RADIO SETS were simple crystal detectors. By 1920 a three element vacuum tube was used in conjunction with a telephone receiver and large antenna for reception of code up to 100 miles or so. Fragile, gassy tubes cost up to six dollars and had short life. By 1923 music broadcasting popularized radio reception. (Adapted from "Robison's Manual of Radio Telegraphy", 1918).

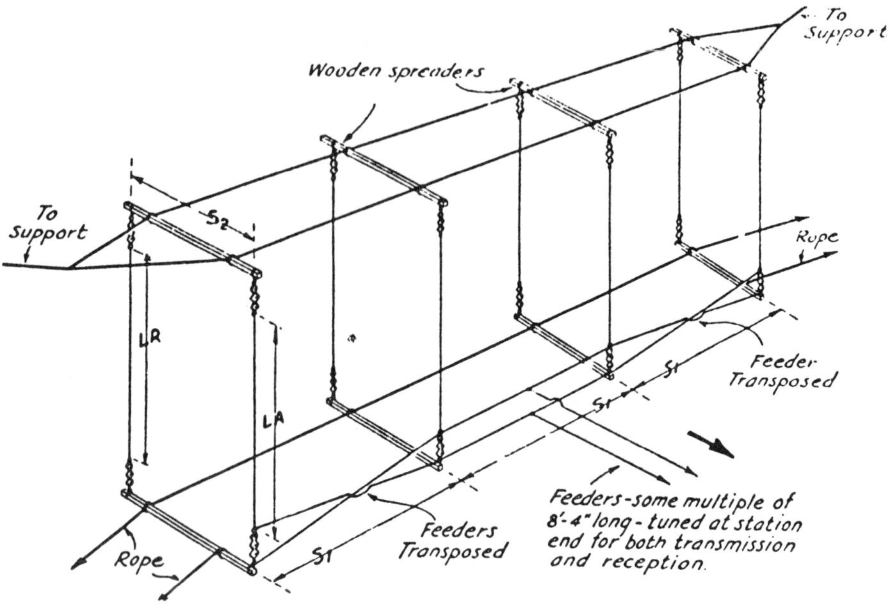

Fig. 4 RECORD SMASHING vhf beam of W1AL (1934). Adapted from an RCA Communications design, this high gain array is constructed of 5/8-inch wood spreaders, #16 wire and light, plastic insulators. Wire is pre-stretched before assembly. For experimenters interested in duplicating this broadband array at 50 MHz, the dimensions will be: LA = 9'4", LR = 9'8", S1 = 9'10", S2 = 4'11". Antenna may be fed with 4:1 balun and coax line. Antenna gain is about 10 dBd. (Drawing from QST, October, 1934).

the sensitive superhetrodyne receiver about 1935, short wave radio transmission became a reality rather than a hit-or-miss proposition. Soon thereafter, the great technological explosion brought about by World War II opened the way for today's communication miracles: radar, television, signals bounced off the moon and Mars, and--finally--the transistor.

All of these modern communication techniques and devices make use of the mysterious "undulations" noted by Faraday so long ago and first put to use by Heinrich Hertz. Of the true nature of the radio wave, however, nothing is known. The radio wave, then, must not be thought of as a thing, but as a way in which things behave. For after a description is given of the behavior of the mysterious radio wave, nothing more can be said, as the ultimate knowledge of the radio wave is a secret locked in the heart of the universe.

Today's radio amateur stands upon the shoulders of giants. Even today, radio is still in a formative phase, and amateur radio in particu-

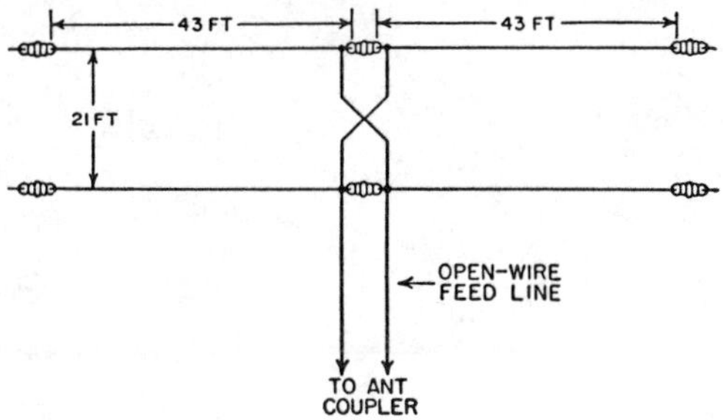

Fig. 5 THE BIDIRECTIONAL ARRAY OF VK2SA (1955). Designed for 14 MHz, the antenna also works well on 7 and 21 MHz. It is supported in the vertical plane by two 41-foot masts. The bottom wires are 20 feet above ground. For experimenters interested in duplicating this beam, ladder line may be used with a balanced antenna tuner. Pattern of beam is at right angles to the plane of the array. Estimated gain is 6 dBd on 14 MHz. Good results should also be obtained on 10 and 18 MHz. (Drawing from QST, October, 1955.)

lar is less than a half-century old. Antennas, and ham antennas to boot, are still very primitive, even though the theory behind them is complex. Today's beam antennas and arrays will probably make the ham of 2001 smile as he shoots off to Paris on a supersonic jet with his space-approved FM transceiver, talking to a satellite repeater as he zips through space!

The Radio Antenna

Heinrich Hertz had a very clear notion of Maxwell's mathematical concept of a radio wave: an interchange of energy in space between free electric and magnetic fields, with the energy radiating outwards from the source as ripples spread out from a stone cast into a quiet pond. The practical problem was to construct a "launching device" that would project the radio wave into space to commence its journey. Today, such a device is called an *antenna* Simply speaking, any structure that radiates and intercepts radio waves is an antenna.

Fig. 6 THIS 85-FOOT ANTENNA is located at a COMSAT earth station and is used to relay high speed communication data between North America and the Far East via the Pacific satellites. This is one of six U.S. earth stations used to handle international communications. Imagine how a comparable dish antenna would work on 2 or 6 meters! Any takers?

Antennas come in all shapes and sizes, from the compact antenna concealed in a transistor radio to the huge towers of the modern television transmitter. Observers have noted rotating disc-shaped radar antennas on boats, backyard television antennas, and the large beam antennas of radio amateurs. Each antenna is shaped and built to do a particular job, but the task of all antennas is to launch and intercept radio waves in the manner most efficient to the task at hand. Hertz, if he were alive today, would understand and appreciate the most complex of modern communication antennas, as the design is based upon laws of nature that are inviolable.

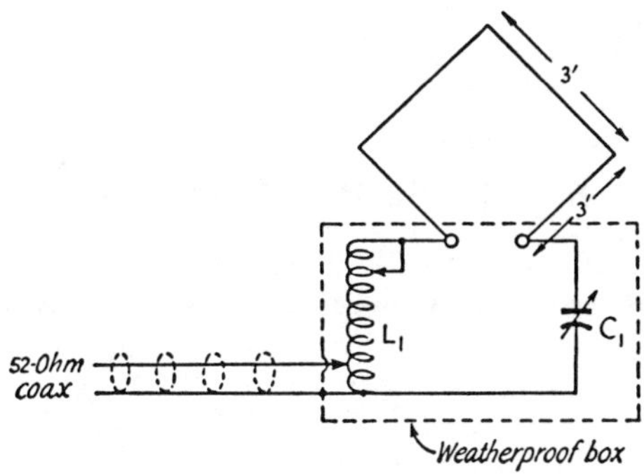

Fig. 7 THE 40 METER TRANSMITTING LOOP OF W4LW (1952) provided 2500 mile contacts. The loop was mounted in the horizontal plane supported by a balcony railing on the second floor. For experimenters interested in duplicating this interesting antenna, the following dimensions are suggested: The loop consists of #10 (or heavier) wire supported on a frame made of 5/8-inch wood dowels. Tuning capacitor is 150 pF. Coil is 2.5 μH spacewound on a 1-inch form. (Strip the insulation from a length of house wiring.) For initial adjustment, coax tap is set at 3 turns and C1 is tuned for resonance, as indicated by SWR meter in line. Tap is adjusted for lowest SWR at resonance. If necessary, adjust L1 by shorting turns. Once correct tap positions are found, the connections are soldered.

Bandwidth of the high-Q loop is about 20 kHz and it must be retuned for change in frequency. For greater bandwidth, it is desirable to make the loop as large as possible, with a corresponding reduction in the inductance of L1. (Drawing from QST, September, 1952.)

Chapter 2

Radio Waves and The Nature of Things

(Old Mother Nature's Mysterious Ways)

The earth and everything on it is continuously bombarded by energy from outer space in the form of waves coming from countless sources. Some waves are useful (sunlight) and some may be dangerous (x-rays). Others are not fully understood (cosmic rays). Many such waves have been recruited in the service of man.

The waves from space arrive helter-skelter but the waves, or groups of waves, are orderly and have their own characteristics. Taken as a whole, when arranged by size, or *wavelength,* they make up the *electromagnetic spectrum* (Figure 1).

All of these waves, constantly in motion about us, vibrate in typical waveform at definite frequencies with the wavelength decreasing as the frequency increases. The number of complete waves per second -- one cycle per second -- is called *Hertz* (Hz).

Wave Bands

About 1880, experimenters developed precise instruments for measuring light waves and by 1900 it was possible to measure the wavelength of radio waves, x-rays and cosmic rays. As the various waves were investigated and measured, they were grouped into *wave bands,* which are groups of waves with similar characteristics, measured in a similar fashion.

The next step was to generate or procure a supply of particular waves for study. Light waves were easy to procure. If a beam of red light was desired, for example, all that was needed was sunlight and a red filter which would block out the unwanted wavelengths and pass only red light. On the other hand, x-rays and radio waves are generated by nature in far-off space, but must be produced on earth as the space waves are too weak to be put to use.

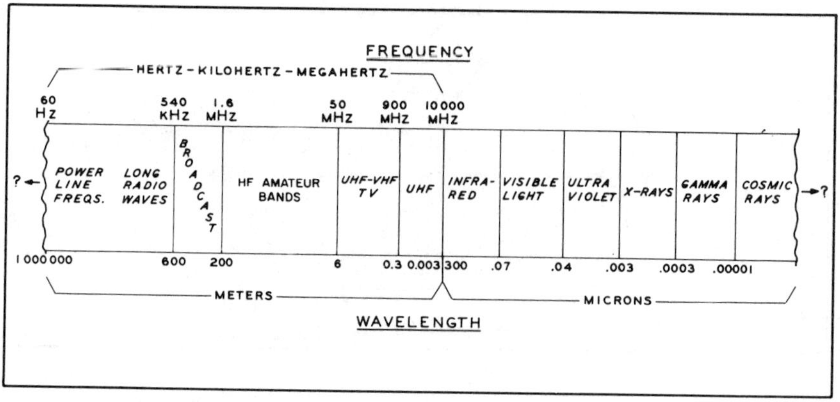

Fig. 1 THE ELECTROMAGNETIC SPECTRUM is a chart showing the relationship between electromagnetic waves arranged by size (wavelength). At the low frequency end of the spectrum are the extremely large radio waves useful for long distance daylight communication. Huge antennas are required to radiate these big waves. The regular broadcast band encompasses medium size waves and still shorter (smaller) waves are useful for long distance, ionospheric-reflected radio communication. The amateur bands fall in this range between 200 and 6 meters. Additional ham bands are in the UHF-VHF region below 6 meters for point-to-point communication.

Gradually, the extremely short radio waves blend into infrared waves, and at still shorter wavelengths, the electromagnetic waves are visible to the eye as light waves. Shorter than light are ultraviolet waves, X-rays, gamma rays and cosmic waves. The size of these tiny waves is expressed in terms of millionths of an inch. Even so, they are identical to the larger radio waves used by radio amateurs.

What waves, if any, exist beyond the ends of the electromagnetic spectrum chart? Scientists suspect that super-long waves may be found in the form of gravitational waves and that micro-miniature waves shorter than cosmic waves exist in the universe. No one is sure about super-long or super-short waves and the puzzle is to find them, as man seeks to learn more about the mysterious universe and all the electromagnetic waves that surround us.

High frequency electric energy can be generated by a radio transmitter and converted into electromagnetic waves by an antenna. The field set up about the antenna is transmitted through space at the speed of light (186,000 miles per second) and may be captured by a second antenna, intercepting the wave and converting it back to electric energy, capable of being detected by a radio receiver.

In other wave bands, the generation and detection of waves is costly, complex and little-known other than by the specialists who make use of the waves. For example, gamma rays are useful for radium therapy,

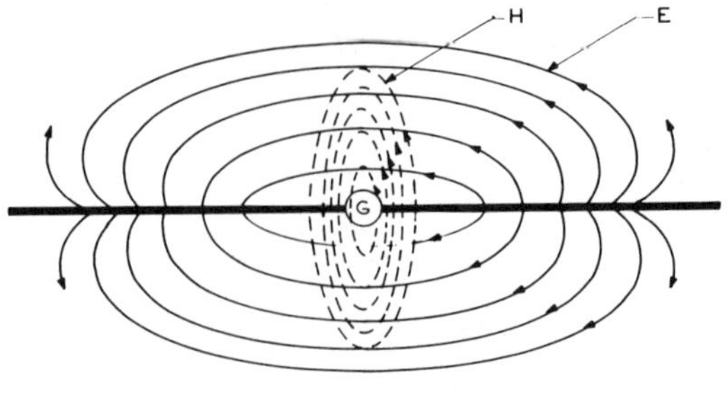

Fig. 2 **MODERN CONCEPT OF A RADIO WAVE** is a combination of magnetic (H) and electric (E) fields set up about an antenna as a result of electric current flowing within the antenna. Energy is transferred back and forth from one field to the other. This action is termed "oscillation". During energy transfer, which can occur millions of times per second, fields may become detatched from the wire and move off into space. A radio signal is apparently made up of the two fields which reinforce each other, with the electromagnetic energy radiating outwards from the antenna. The sum of the two fields is called an "electromagnetic field". If the field cuts another conductor (antenna), some of the energy in the field will set electrons in motion in the conductor. The electron movement (current) may be detected by a radio receiver.

x-rays for medical and industrial applications, ultra violet waves for lasers and black light, infra red waves for drying and photography, microwaves for radar--to name a few.

Of all the wave bands, however, the radio waves have received the most use. This Handbook covers their launching and interception by antennas, and how this knowledge can best be put to use by amateur radio operators.

The Two-Way Antenna

It was discovered early in the game that an antenna acts in the same fashion whether it is transmitting or receiving energy. Thus, the general characteristics of the antenna are the same in either instance. For simplicity, then, we can examine or discuss a particular antenna in terms of transmission or reception, with the assurance that the observations noted in one case apply to the other. The basic theory regarding similar receiving and transmitting characteristics is called *reciprocity* and, as applied to antennas, was first stated by Lord Rayleigh of England in 1877, long before radio transmission was generally known. His basic theory is still true today.

Antenna Resonance

Electrical energy is radiated into space and retrieved from space by an antenna. An antenna is a length of metal that conducts electricity and is so located that it is surrounded by space. High frequency electric energy flowing in the antenna sets up an electromagnetic field about the conductor which expands into space at the speed of light. Conversely, an electromagnetic field meeting an antenna sets up a high frequency electric current within the antenna. Thus, *any* conductor of electricity functions as an emitter or receptor of electromagnetic waves, with a varying degree of efficiency(Figure 2).

For highest efficiency, the antenna must bear some relationship to the length of the radio wave. In the early days of radio, wire fences were often used for reception and crude "cage" antennas of many wires were used for transmission without any real concept of trimming or

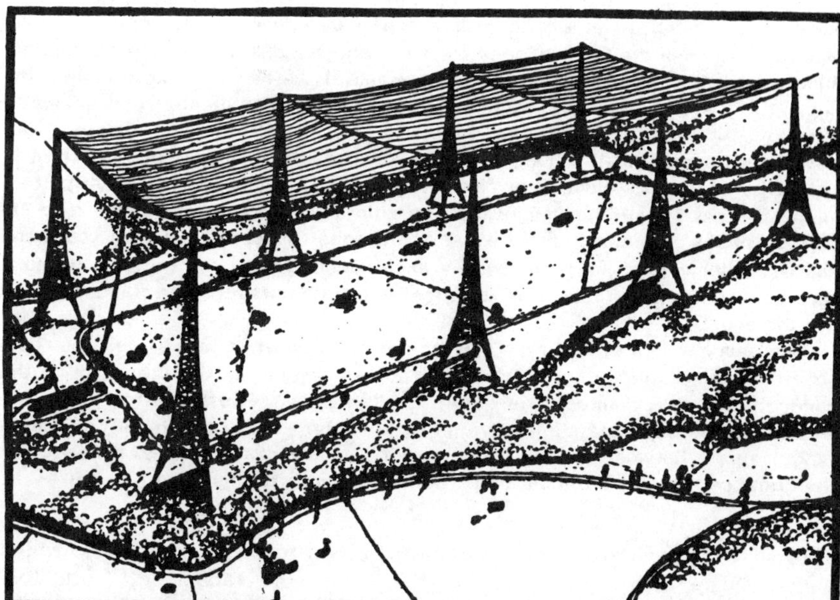

Fig. 3 WORLD'S BIGGEST ANTENNA? Huge antenna built in Lafayette, France, in 1917 for super-power long-wave station (which was never finished). The antenna covered several hundred acres and was supported on 600 foot high towers. Early radio books give vivid descriptions of big radio antennas, but operation and theory was often vague and misleading. (Adapted from "Radio Theory and Operating", by Loomis, 1925).

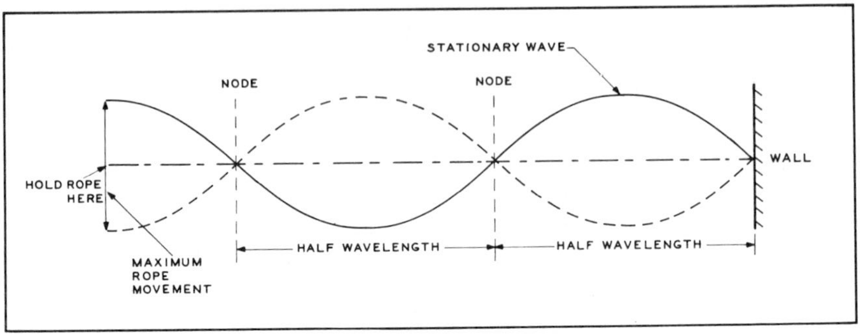

Fig. 4 RADIO WAVELENGTH. A simple case of wavelength brought about by the reflection of a wave can be observed when a rope is given a quick series of flips. A wave travels along the rope until it reaches the far end, from where it travels back along the rope to your hand, continuing back and forth in this manner until the motion dies out. By flipping the rope in the proper sequence, you can make the wave continue to run back and forth along the rope.

With a little practice, you can flip the rope in the proper sequence and a succession of waves at equal intervals will travel along the rope. When reflected back from the far end, they meet the oncoming waves whose lengths are equal to those waves coming from the far end of the rope. At some points, the conflicting waves reinforce each other and at other points the rope seems to not move at all. Points of zero movement are found along the rope one-half wavelength apart; at all other points the rope vibrates. The vibratory pattern is called a "stationary wave", or "standing wave", as the overall wave pattern moves neither forward nor backward. The points of no movement of the rope are called "nodes". The stationary wave on the rope is trapped between your hand and the other end of the rope, and by experimenting, you can get various numbers of standing wave nodes on the rope, depending upon the rate at which you flip the end.

Stationary waves of this type may be set up in an electrical circuit, or along an electrical conductor by electrical impulses applied to the circuit or to the conductor. Such a conductor is called an "antenna". Shown in this picture is an antenna having a standing wave on it, with three nodal points. A half-wavelength exists between any two nodal points. In the case of an 80 meter wave, the distance between adjacent nodal points is about 135 feet.

otherwise adjusting the length of the antenna to the length of the radio wave. "The bigger the antenna, the better the reception," was the motto. Such antennas, of course, were very inefficient (Figure 3).

A great body of literature exists today attesting to the many varied and important contributions made by numerous experimenters in the quest for better and more efficient antenna systems. Early in the game, it was found that an antenna of a certain length did not exhibit the same characteristics at all frequencies. Best results were obtained when the antenna was adjusted physically or electrically to be in *proportion to the length of the radio wave* . If the antenna was shorter than the wave, extra wire could be added in the form of a coil to make the antenna electrically *longer*. If the antenna was too long, a capacitor could be added in series with the antenna to make it electrically *shorter*. The antenna and its auxiliary *tuner* could thus be adjusted to the condition of *resonance*, or electrical compatibility with the radio wave in use (Figure 4).

Resonance and Antenna Length

For any antenna there is one frequency, called the *resonant frequency*, at which various characteristics of the antenna are in a state of electrical balance, and at which frequency the antenna is in a condition of maximum efficiency. The resonant frequency is a function of the *electrical length* of the antenna, which may or may not bear a relationship to the physical length in feet and inches. Any antenna may be tuned to resonance by auxiliary gadgets, but such devices may be a nuisance and of questionable efficiency. A resonant antenna requires no such devices and is a simple and effective radiator and receiver of radio energy. The length of the radio wave and the antenna is expressed in terms of *wavelength*, and that term is directly related to the *frequency* of the radio wave, as we shall see shortly.

Wavelength and Frequency

Radio waves exist because it takes a certain amount of time for electrical energy to travel from point to point. When a pebble is dropped into water, the resulting disturbance does not reach the edge of the pool immediately. Rather, a wave of water starts out from the place where the pebble hits and proceeds towards the edge at a definite speed. Electrical energy normally travels at 186,000 miles per second, or 300,000,000 meters per second. Thus, if a radio antenna is emitting a pulse of radio energy at a rate of 14,000,000 pulses per second (14,000 KHz -- the 20 meter band), one pulse will travel about 20 meters before another pulse is emitted. At 186,000 pulses per second, on the other hand, the distance between pulses is about 5000 feet. At 3,000,000,000 pulses per second (3,000 gigacycles), the distance between pulses is only about 4 inches. This corresponds to a radar signal. Thus, exactly as in the case of the pebble dropped into water, a mathematical relationship exists between the distance a radio pulse will travel during the time required for one pulse (one cycle) to occur, as shown in Figure 5.

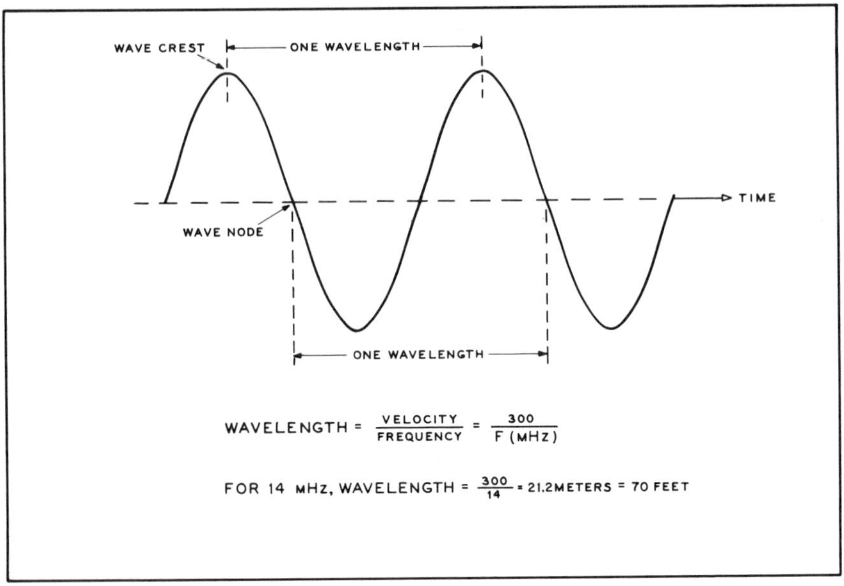

Fig. 5 WAVELENGTH AND FREQUENCY are related to each other by a simple formula, as explained in the text. The wavelength is the distance between two wave crests (or between two wave nodes). The velocity (speed) of the electromagnetic wave is 300,000,000 meters per second, the same as the speed of light. The frequency, or repetition rate, of the wave is the number of waves which pass a given point in one second. Wavelength is measured in units called "meters", one meter being equal to 3.28 feet. The frequency is measured in "cycles per second", or units called "Hertz". Thus, one cycle per second is one Hertz.

Since wavelength and frequency have an inverse relationship, it follows that long waves have a low frequency and short waves have a high frequency.

The length of any electromagnetic wave can be found by the formula:

$$\text{Wavelength in meters} = \frac{300{,}000{,}000}{\text{frequency in cycles per second}}$$

In the case of high frequency radio, the wavelength is expressed more conveniently as:

$$\text{Wavelength in meters} = \frac{300}{\text{frequency in megahertz (MHz)}}$$

The illustration tells the story better than words.

Measurements in the United States are usually expressed in feet and inches rather than in meters (and centimeters). This formula, converted into more familiar terms, then, is:

$$\text{One wavelength, in feet} = \frac{984}{\text{frequency in megahertz (MHz)}}$$

A half-wavelength is determined by dividing the formula by 2, thus:

$$\text{One-half wavelength, in feet} = \frac{492}{\text{frequency in megahertz (MHz)}}$$

This is the fundamental formula from which many significant lengths in antenna work are developed and is defined as the *electrical length* of a half-wave antenna element, when no factors exist that modify the speed of the radio wave. In real-life situations, many factors exist that alter the *physical length* of a half-wave antenna, or any conductor used for an antenna. If this were not true, there would be no need for this Handbook!

End effect and *thickness* of the antenna element must be taken into account when determining the actual physical length of any antenna element. The end effect is caused by the presence of insulators or other material that may be used to support the antenna at the ends and also by the abrupt transition from conductor to the surrounding atmosphere at this critical point.

Since practical antennas have thickness as well as length, the actual length departs to a small degree from the electrical length. Generally speaking, the larger in diameter the antenna element, the greater the departure from the electrical wavelength. Thus, for a given frequency an antenna having a small cross-sectional area in the conductor will be longer physically than an antenna having a larger cross-sectional area.

Very Short Antennas

The physical length of an antenna often poses a problem, especially in portable or mobile work in an automobile or boat. It is possible to use a very short antenna (at some sacrifice in overall efficiency) and electrically "load" it so it appears as an antenna of greater length. The missing portion of the antenna takes the form of a *loading coil* of wire wound on an insulated form (or supported by its own rigidity). The coil, for simplicity's sake, can be thought of as the missing antenna length, wound up into a compact structure. The coil may be placed at

CUBICAL QUAD ANTENNA is popular among radio amateur DXers. Quad beam delivers the most signal gain per area of any inexpensive beam antenna known, although Quad assembly is tricky and complicated.

one end of the antenna, or along the length of the antenna. You see these loading coils as bulges at the base of the antenna, or along its length. Coil design and construction is crucial, and the efficiency of the antenna depends upon the skill of design and manufacturing, and the expertise of the person making the antenna installation. It's not a game for beginners. In any event, the portion of the antenna used up in the coil contributes nothing to the radiation prowess of the antenna. Usually, the smaller the antenna in relation to the operating wavelength, the bigger the loading coil, the lower the antenna efficiency, and the weaker your signal! *There is no substitute for a full size, resonant antenna!*

Summary

The radio wave is an electric "undulation" which is a portion of the electromagnetic spectrum. The radio wave is specified in terms of size (wavelength) and frequency (cycles per second). It may be radiated and intercepted by an antenna which converts radio waves into high frequency electrical energy and vice-versa. The antenna has the same characteristics when transmitting or receiving and has the greatest efficiency when it is in a state of resonance.

Chapter 3

Your Antenna and Signal Interception

(How Does the Radio Wave Get Down That Skinny Cable?)

What happens when the radio wave meets the antenna? Why does a beam antenna make your signal louder? What is the meaning of antenna gain? Where does the gain come from? Good questions, and ones not easily answered.

To start with, a radio wave travels unhindered through space until it meets a conductor. In our case, the conductor is the antenna, and it is made of metal. By definition, a metallic object is one that has "free" electrons able to move about within the conductor. The intercepted radio wave imparts energy (motion) to these electrons, which move in a direction corresponding to the direction of the wave--along the conductor.

Inside The Antenna

The electron movement in the antenna element actually *reradiates a portion of the radio wave back into space* while the remainder is captured by the radio receiver attached to the antenna. This action takes place regardless of the length of the antenna and is shown in Figure 1.

When a radio receiver is attached to an antenna which is energized by a radio wave, a portion of power is extracted from the antenna and is passed to the receiver circuits. The total electron current flowing in the antenna (while very feeble) may be thought of as the sum of many individual currents, all acting along the length of the antenna. When all of the individual currents add up at the receiver, the maximum possible value of current is extracted from the antenna. This condition of maximum current is our old friend, antenna resonance. Resonance is established when the antenna bears a certain relationship to the length of the intercepted radio wave, usually found at multiples of one-half wavelength. The basic half-wavelength antenna is called a *dipole* antenna. At 20 meters, for example, a dipole is about 33 feet long.

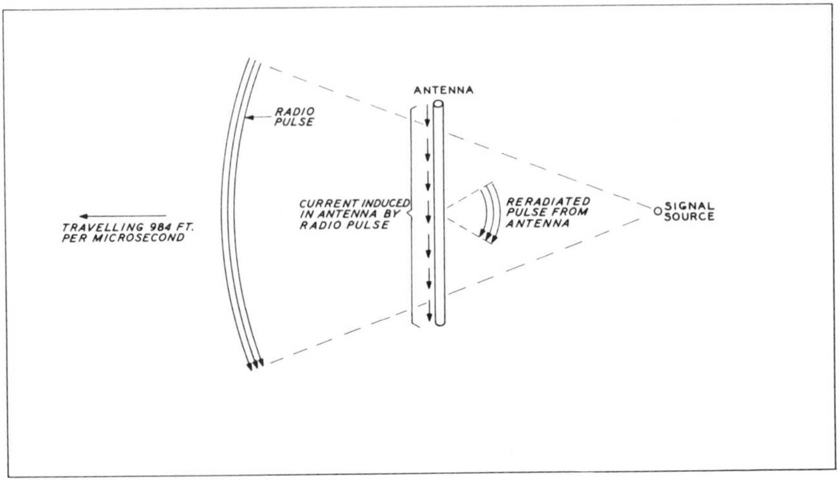

Fig. 1 RADIO WAVE MEETS ANTENNA. At the right is a radio signal source, such as your transmitter, radiating a radio wave moving to the left. For this example, the wave is just a short burst (''pulse'') of energy. When the pulse meets the antenna (center), it causes free electrons to flow along the antenna in the form of an electric current. This current re-radiates a portion of the intercepted radio wave back into space. The original radio pulse passes to the left, beyond the antenna, followed shortly by the re-radiated energy from the antenna, which spreads out both to the right and left of the antenna. Now we have two pulses: the original one and a weaker, re-radiated one following along a fraction of a second later. When the current flowing in the antenna is a maximum value, the antenna is said to be ''resonant''.

Power Transfer

With any type of electrical power source, whether it be a storage battery, a generator or a radio transmitter, maximum power is delivered from the source to the destination *(load)* when the load and source each have equal voltage and equal current at their terminals. In this condition, the source *(radio transmitter)* and load *(antenna)* are said to be *matched*. Specifically, to obtain the maximum transfer of power from a radio transmitter to an antenna, or from an antenna to a receiver, the antenna must meet two requirements:

1 - The antenna must be resonant at the frequency of the radio wave so as to deliver a maximum signal current to the receiver, or to extract maximum power from the transmitter.

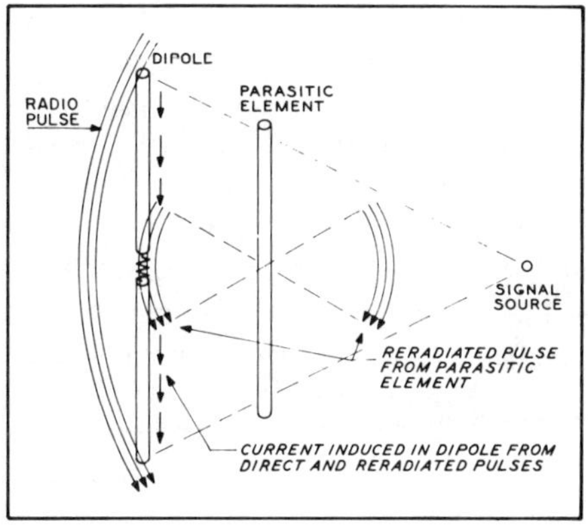

Figure 2

RADIO ANTENNA connected to load, which could be your receiver. When pulse meets parasitic, most of the radio energy is radiated back into space to loaded antenna. If phasing (timing) is correct, energy from parasitic element reinforces current flowing in loaded antenna. Yagi beams are made up of loaded antenna plus one or more parasitic elements.

2- The signal voltage and current in the antenna and the receiver (or transmitter) must be matched to achieve best transfer of energy.

Even when these two demands are met, the matched, resonant antenna delivers only a portion of the current extracted from the radio wave to the receiver, reradiating the remaining portion of the current back into space a fraction of a second after the wave has arrived at the antenna. As you'll see, this reradiated wave can be made to perform in a very useful manner.

The Beam Antenna and Signal Gain

A *beam antenna* is an antenna that concentrates radio energy in a given direction at the expense of radiation in other directions. A good analogy is the headlight of an automobile which concentrates the light into a single beam, at the expense of illumination to the sides and back. A beam antenna concentrates the radio wave in a similar fashion, although the beam of an antenna is not nearly so well defined as that of a headlight beam.

A resonant antenna reradiates virtually all of the intercepted radio wave if it is not connected to a load. When such a device is placed near another antenna connected to a radio receiver, the unconnected antenna is called a *parasitic element*. Figure 2 shows a parasitic element placed near an antenna which is connected to a load. The parasitic element is placed between the antenna and the signal source. The antenna is now exposed to the direct radio wave from the source and also to the reradiated radio wave from the parasitic element, which reaches the antenna a fraction of a second later.

If the reradiated wave from the parasitic element reaches the antenna at the proper time, *it reinforces the radio wave received directly from the source by the antenna.* Adjusting the length of the parasitic element and its spacing from the antenna accomplishes this. Such a parasitic element is called a *director.*

When the parasitic element is placed *behind* the antenna and the signal source, and is adjusted to enhance the received signal, it is called a *reflector*. Beam antennas are made up of a connected, or "active" antenna, working in conjunction with a single reflector and one or more directors.

Antenna Power Gain

Power Gain (or signal gain) is a term used to express the power increase in receiving or transmitting of one antenna as compared to a a standard comparison antenna. The comparison antenna may be either a dipole, or it may be an imaginary device called an *isotropic antenna,* useful for mathematical computations in the laboratory. The isotropic antenna is considered to be so small that it radiates equally well in all directions--something that no actual antenna can accomplish.

Of great interest to the prospective purchaser or user of a beam antenna is the amount of power that still escapes from the sides and back of the array. The ratio of power radiated in the forward direction of the beam as compared to the amount radiated in the opposite (back) direction is called the *front-to-back ratio* of the antenna.

Beam antennas come in all shapes and sizes and prove to be a willing subject for imaginative manufacturers to exploit, sometimes at the expense of the buyer, dazzled by important sounding words and inflated claims. Shown in this Handbook, however, are some "honest" beams that are inexpensive and work well. Before this, it is well to investigate and discuss the "yardstick" by which antennas of all kinds are measured. This "yardstick," when properly interpreted, can give you a lot of useful information about any antenna. The "yardstick" is called the *decibel.*

The Mysterious Decibel - The Unit of Truth!

Power gain and front-to-back ratio performance of beam antennas are expressed in terms of decibels (abbreviated *db*). The decibel is not a unit of power, but a *ratio of power levels.* In antenna practice, the decibel may be used as an absolute unit of measure by fixing an arbitrary level of reference. If the reference level is a dipole antenna

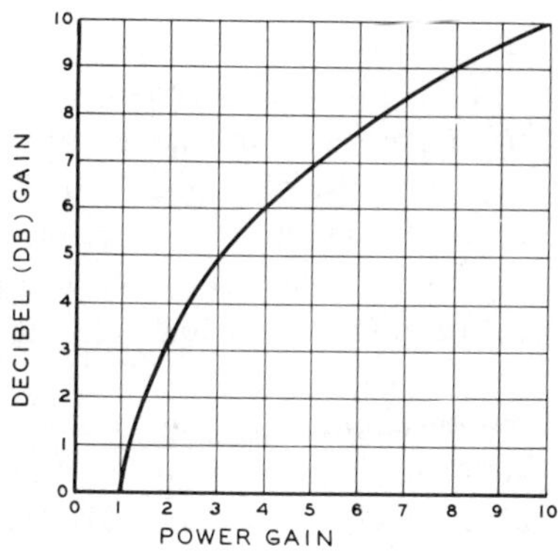

POWER GAIN OVER ONE WATT (UNITY)	DECIBEL GAIN
1.00	0.0
1.12	0.5
1.26	1.0
1.42	1.5
1.58	2.0
2.00	3.0
2.51	4.0
3.16	5.0
3.98	6.0
6.31	8.0
10.00	10.0

Fig. 3 THE DECIBEL--THE "YARDSTICK" OF PERFORMANCE. The decibel is the "yardstick" of performance in the electronics world just as "horsepower" is the yardstick in the automotive world. The decibel unit expresses the ratio between two power levels in an electrical circuit. Originally, the decibel measured the loss of voice power over one mile of telephone wire.

This chart shows the relationship between the decibel and the watt for gains of up to 10 over a reference level of one watt. For example, a power increase of 6.31 represents a decibel gain of 8. A decibel gain of 3 represents a power gain of 2.

Antenna gain is commonly expressed in decibels gain over a reference source, such as an isotropic radiator or a dipole. It is important to know the reference as the decibel is a ratio of change of power and can only be used as an absolute unit by fixing a level of reference. If a dipole is chosen as a reference antenna, it is said to have a power gain of unity (one).

(for example), another antenna may be said to have "so-many decibels power gain over the reference dipole."

Antenna gain expressed in decibels without mention of reference level is meaningless. It is often deliberately expressed in this fashion by some antenna manufacturers who hope to overawe the reader into thinking he is getting more power gain for his money than he actually is. For example, the statement, "Ten decibels power gain!!!" doesn't mean a thing, because it does not reference the gain to any level of measurement. If the statement said, "Ten decibels power gain over reference dipole" the statement has meaning, as the level of reference is given. Beware expressions of antenna gain that are unreferenced!

Shown in Figure 3 is a decibel chart and table giving the relationship between decibels and power. A power ratio of 4 provides a change of 6 decibels. A power ratio of 10 provides a change of 10 decibels. This relationship holds true regardless of the electrical circuit, and the decibel is commonly used in all electrical work throughout the world. It is a useful form of "shorthand" understood by all electrical engineers, regardless of whether they work with radio, audio circuits, power lines, or space communication. Like it or not, the decibel is here to stay.

A useful rule-of-thumb is that a power change of one decibel is *just noticeable* on a quiet communication circuit, and a power change of 3 decibels is about the smallest change that is economically feasible. That is to say, a power increase of less than 3 decibels is not worthwhile in most instances, as it is marginally helpful, and a power decrease of 3 decibels is not especially harmful.

Translated into watts, this means that if your transmitter has a power output of 100 watts, boosting it to 150 watts by careful tuning won't be of much help! Or, if your power should drop from 150 watts to 100 watts, the operator at the other end of your circuit probably won't notice any difference in signal strength at all!

Doubling power provides a boost of 3 decibels, and that's about the least change that can really be noticed over a period of time. Accordingly, an antenna that provides 3 decibels power gain over a dipole could be worthwhile.

Experience has shown that a power increase of 5 decibels or so is definitely worthwhile, and a power change of 10 or more decibels is often startling in the results it achieves over a difficult communication path.

The Curious Antenna

The concept of the antenna -- a simple length of aluminum tubing or copper wire -- taking electrical energy and transforming it into invisible radio waves that cross space at the speed of light is a breathtaking idea. Textbooks of great weight and complexity have been written about antennas, and very large antenna arrays have been built that defy comprehension by the layman. This Handbook ignores complexity and covers in detail the popular and everyday antennas encountered in ham radio and discusses their operation in simple, nontechnical terms you can understand.

You, the reader, will not be placed in the position of the little girl sent to the encyclopedia to learn about penguins. When she finished reading, her mother asked if she was satisfied with what she had read. The girl replied, "Not really, the encyclopedia told me more about penguins than I wanted to know!"

This chapter (much shorter than an encyclopedia) covers properties of the antenna that are important to you. No complicated mathematics at all! Let's start with antenna polarization, as shown in Figure 4.

Antenna Polarization

Polarization of a radio wave is determined by the position and direction of the electric field with respect to the earth's surface. If the lines of the electric field of a radio wave are parallel to the ground, the wave is *horizontally polarized*. If the lines are perpendicular to the ground, the wave is *vertically* polarized. Since you can't see the field, and it takes instruments to determine the polarity, it's sufficient to know that, for the majority of ham antennas, the electric field is in the same plane as the antenna element so that a vertical antenna is vertically polarized and a horizontal antenna is horizontally polarized. For high frequency communication polarization is not particularly important as ionospheric-reflected waves tend to loose their initial polarization. It should be mentioned, however, that vertical receiving antennas tend to pick up more man-made noise and interference than do their horizontally polarized counterparts.

Antenna Operation

While you can't do very much about the tricks the ionosphere may play on you, there's plenty you can do about having a good, efficient antenna system. Let's turn, therefore, from radio propagation to the antenna itself and see what goes on inside this interesting device. The remainder of this chapter discusses some of the "ground rules" of antenna operation.

Impedance --- More Power To You!

You'll often see terms such as *surge impedance, feed impedance, radiation resistance, input impedance* and the like, used with reference to antennas. These terms are telling you something you should know about the antenna. The big question is: what's it all about?

The simplest answer is that when the complete antenna system is

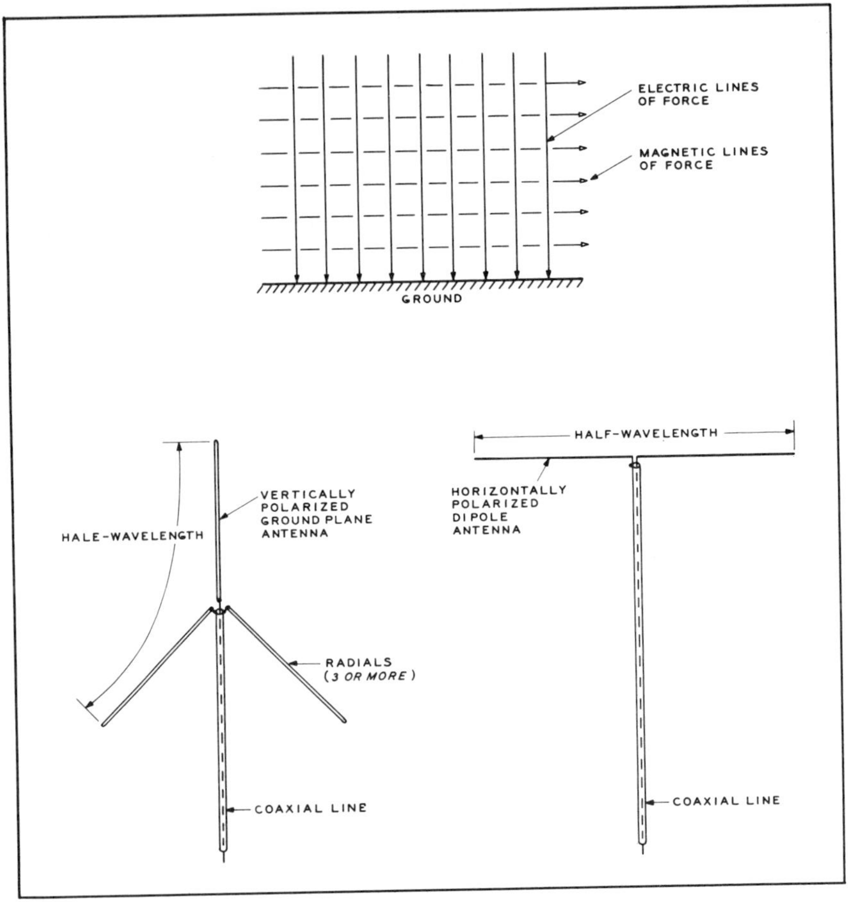

Fig. 4 CAN YOU SEE A RADIO WAVE? If you could, perhaps this is what you would see if a radio wave was travelling out of the page towards you! It is a fanciful drawing of a radio wave showing the electric and magnetic fields. Electric fields are commonly represented by solid lines and magnetic fields by dashed lines. The arrow heads indicate the direction of the field. In this illustration, the lines of the magnetic field are parallel to the ground and the lines of the electric field are vertical. According to the rules of the game, therefore, this wave is vertically polarized. For common antennas, the electric field is in the same plane as the antenna element so this represents the radio wave leaving a vertical antenna, such as a ground plane. Of course, the field surrounds the antenna, so this picture represents just a small portion of the wave. The magnetic field actually surrounds the antenna and the magnetic lines move outward, just as ripples in a pond spread when a stone is cast in the water. The electromagnetic field concept is a difficult one to grasp, so don't loose too much sleep over antenna polarization.

in a state of electrical balance (the state of balance being defined by these various terms), the overall operation of the system is at the highest possible efficiency. This means the best reception and the strongest signals.

Many readers of this Handbook are familiar with Ohm's Law, which is:

$$R = \frac{E}{I}$$

Simply, this formula tells you that the resistance of any electrical circuit (R) is proportional to the *ratio* of the voltage (E) to the current (I) in the circuit. This is *not* the definition of Ohm's Law in most books, which usually ends up talking about resistance in terms of ohms, and in terms of physical resistors. As far as antennas go, the idea of *resistance*, as such, is misleading, and it is much better to think of the word "resistance" as expressing the ratio of the voltage to the current at any given point in the antenna circuit. Thus at a high "resistance" point, the ratio of voltage to current is high and at a low "resistance" point, the ratio of voltage to current is low.

Remember in Chapter I the short discussion of Maxwell's discovery of the interplay of energy between the electric and magnetic fields *around* a wire, and the current flowing *within* the wire? Ohm's Law, in the broadest sense, applies in this case, describing the ratio between the electric field and the current flow in terms of a fictitious resistance, which may be termed the *radiation resistance* of the wire, or antenna:

$$\text{Radiation Resistance of the Antenna} = \frac{\text{Voltage field about the antenna (E)}}{\text{Current flowing within the antenna (I)}}$$

or, in Mr. Ohm's terms: $R = \frac{E}{I}$, which is the same law that applies to direct current circuits, car batteries, your refrigerator motor, the TV set, and such.

Radiation Resistance

To repeat: the terms radiation resistance, input impedance, surge impedance, feed impedance, etc., all refer to the *ratio* between the voltage field about a conductor and the current flowing within the conductor. The expression of this ratio is in terms of ohms. This term is, perhaps, unfortunate, as it implies some kind of actual resistance within the antenna, which does not really exist. As a dodge, this "resistance" is referred to as radiation resistance, which is OK as long as you understand that it means a *ratio of voltage to current*, and

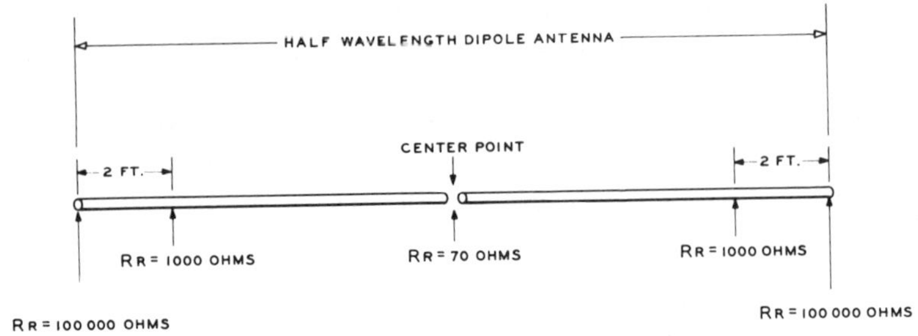

Fig. 5 WHERE DOES THE POWER IN AN ANTENNA GO? It has to go someplace, and it does. It is radiated into space in the same manner as heat and light. It leaves the antenna, never to return. It is difficult to conceive of radio power vanishing into empty space, so a fictituous term (radiation resistance) is used to describe the radiating effectiveness of the antenna. Radiation resistance, as explained in the text, is an expression of the ratio between the voltage field of the antenna and the current in the antenna at any given point along the antenna. In a simple dipole, the radiation resistance at the center is quite low, increasing towards the ends of the antenna. Thus, the voltage to current ratio is different at different points along the antenna. Most antennas are fed at the center and -- in a dipole -- this is the point of lowest radiation resistance. The dipole in this illustration is suspended in space about a half-wavelength above the ground, and the measurements shown may actually be found when the proper instruments are used. Moving the antenna closer to, or away from the ground, will alter all the measurements drastically.

not an actual value of resistance.

Let's take an example: Suppose 3 watts of radio frequency (r-f) power is applied to the center of a dipole antenna. With appropriate (and expensive) instruments, it is determined that the current flowing in the dipole at this point is 0.21 ampere and the voltage field about the dipole is 14.7 volts (Figure 5). The radiation resistance (R_R) is, therefore:

$$\text{Radiation Resistance} = \frac{E}{I} = \frac{14.7}{0.21} = 70 \text{ ohms}$$

No 70 ohm resistance exists in the antenna. But, if the antenna were removed, and a 70 ohm resistor substituted in its place, the *same ratio* of voltage to current would exist, and the same absolute values of voltage and current.

Now let's move out along the dipole's length and measure the voltage and current towards one end, say about two feet from the end of a dipole. What do we find? The same amount of power exists at this point (3 watts) but the voltage field has increased to 55 volts and the current in the dipole has dropped to .055 amperes. The fictitious radiation resistance value is now expressed by:

$$\text{Radiation Resistance} = \frac{E}{I} = \frac{55}{.055} = 1000 \text{ ohms}$$

Finally, let's examine conditions at the very tip of the antenna. Logic tells us that current is zero (since it has no place to go) and voltage should therefore be infinite, if Ohm's Law is obeyed. Practically, this is not true, as the current can "leak" into the atmosphere, or into the insulators that support the dipole. The same 3 watts exist, but the instruments show that the voltage field is now 550 volts and the current is .0055 amperes. The radiation resistance, accordingly, is:

$$\text{Radiation Resistance} = \frac{E}{I} = \frac{550}{.0055} = 100{,}000 \text{ ohms}$$

So! While the power at any point in the antenna is constant, the voltage to current ratio varies all along the antenna, and the radiation resistance (or whatever you want to call this ratio) varies too. The fact of interest to the antenna designer is the ratio *at the point at which he feeds energy to the antenna* by attaching a transmission line. This is usually at the center point, since the radiation resistance ratio is reasonably low and manageable. If the radiation resistance is too low or too high, other problems arise in feeding energy to the antenna which are beyond the scope of this Handbook. Suffice to say that radiation resistance values of 25 to 100 ohms or so are a "target" range for antenna feed points.

Well, What Does It All Mean?

In summary, then, the various terms mentioned previously expressing "radiation resistance" are *not* expressions of actual resistance itself, but an expression of a ratio between the voltage field of the antenna and the current within the antenna at any given point along the antenna. If you can forget the traditional concept of resistance and think instead of the *ratio of voltage to current,* you can more readily understand some of the more obscure representations of working antennas and the over-complex and confusing terms that express these concepts.

Impedance And All That!

So far we've looked at radiation resistance. What about impedance, another OK-term bandied about by the knowledgeable to impress the natives? Remember---the term *resistance* is used in antenna work in its broader interpretation as the voltage to current ratio at which power is consumed or transferred. The antenna, of course, does not consume power, it merely transfers it into the space about it. Nevertheless, for the purpose of circuit design it is convenient to substitute the resistance concept because the circuit *behavior* conforms to that of an actual, real-life resistance.

The term *impedance* is used in a comparable sense, too. It is also a voltage to current ratio, but it expresses an even more general concept than resistance because it implies that all of the power supplied to the circuit *may not be consumed or passed on,* but a certain proportion of it can be returned to the source during some part of the radio energy cycle. The proportion of power consumed or passed on is expressed in terms of *radiation resistance* and the portion of the power returned to the source is expressed in terms of *reactance*. Reactance is undesirable because it makes the antenna difficult to "load" (take power from) the transmitter.

Purists draw a distinct difference between resistance and reactance; however, in antenna talk, the terms are often interchanged. This can be done in the case of a resonant antenna without stretching the truth. And in the case of a ham antenna, operating on various frequencies in one band, the terms can be interchanged, even though there is a distinction between radiation resistance and impedance or reactance that will be more easily understood when we tackle the problem of transmission lines and standing wave ratio later in this Handbook.

Now, we have a vague idea of radiation resistance and the various terms that mean the same thing, and we know that the voltage and current ratio is different at different points along the antenna. A very nice picture can now be drawn of the variations, as explained in the next section.

Antenna Voltage and Current

Now look at the drawing of Figure 6. This shows two antennas, one a full wavelength antenna and the other a half-wavelength antenna. The "radio wave" can be drawn on these figures and now, based upon the previous discussion, it can be seen that this is a current wave, hav-

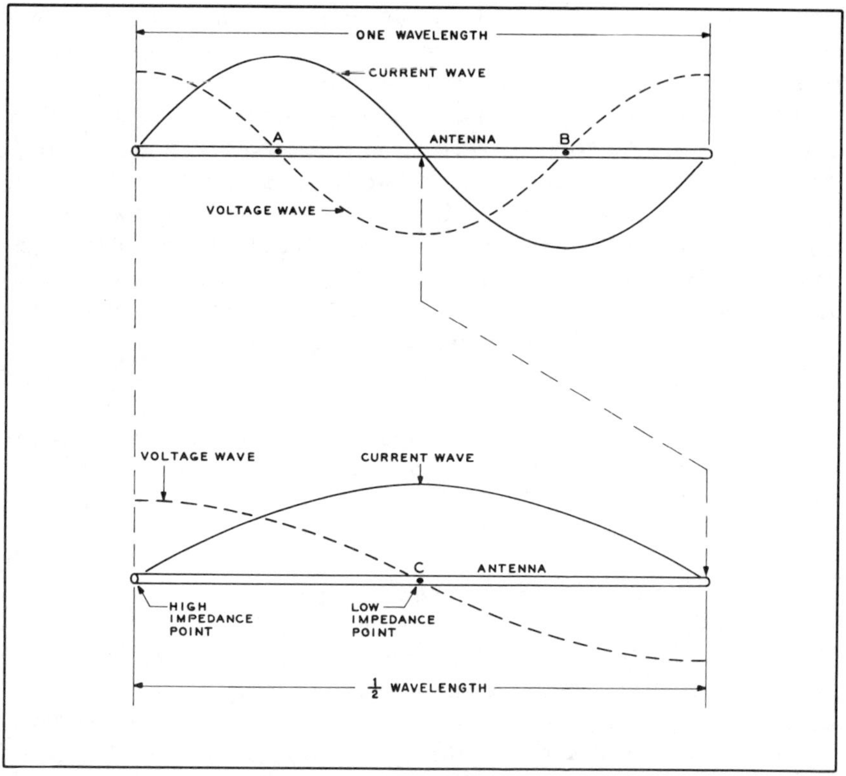

Fig. 6 DID YOU KNOW YOUR ANTENNA LOOKS LIKE THIS? Shown at the top is a one wavelength long antenna for 10 meters. At the bottom is a half-wavelength antenna. The voltage and current waves for a radio signal are drawn on the antennas. The height of the wave indicates the strength ("amplitude") at that point on the antenna. The voltage wave (dashed line) is maximum at the high impedance points on both antennas, and is a minimum at the low impedance points. The current wave (solid line) is exactly opposite, being a minimum at the ends of the antenna (the high impedance points) and a maximum at the low impedance points. In the case of the half-wavelength antenna, this is at the center.

You don't believe it? Well, if sufficient radio power was applied to either antenna (100 watts, or so), the voltage wave could be traced out by holding a neon lamp or flourescent tube near the antenna. The lamp will glow brightly from induced power at the high impedance points and will be extinguished at low impedance points. The lamp need not touch the antenna for this experiment.

For convenience, ham antennas are usually fed at low impedance points, as low impedance coax transmission line is readily available at modest cost. In the case of the one wavelength antenna, the feedline would be attached at either point A or point B. The feedline is attached at point C for the half-wavelength antenna.

ing a minimum value at the ends of the antenna and a maximum value -- in the case of the half-wave antenna --at the middle. Also observe that the one wavelength antenna has two current maximums on it.

Figure 6 can now be modified to show the voltage wave as well as the current wave for both antennas. The voltage wave is a dashed line. The voltage wave is maximum at the high resistance ratio (impedance) ends of the antenna and the voltage curve is exactly opposite to the current curve. That is, the crest of one falls at the low point of the other. Radio lingo says these waves are "out of phase". Note, too, that *voltage is always maximum at the end of the antenna,* and a minimum one-quarter wavelength along the antenna from the end. Likewise, *current is always minimum at the end of the antenna* and a maximum one-quarter wavelength along the antenna from the end. Finally, it should be noted that energy is usually (not always) introduced into the antenna at a point of maximum current (at the crest of the current wave). Thus, the dipole antenna is fed at the center--- the point of highest current.

Upwards And Onwards

Now that you have an idea of some of the terms and concepts used in antenna work, and a bit of the interesting theory behind the development of today's antennas, let's examine some antennas and feedlines. Antennas are like blondes: -- "There are only two types of blondes --- good and better." So, let's look at some of the better ones (antennas, that is).

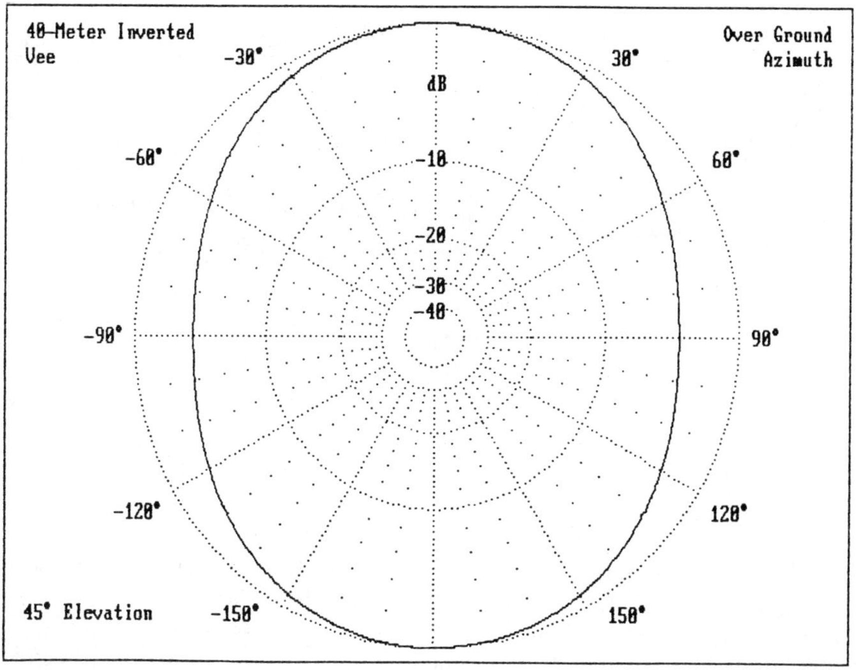

Fig. 7 When a dipole is bent to form an inverted-V, the pattern nulls are partially filled in by vertically polarized radiation. The V angle can be adjusted to vary the omnidirectional coverage. Pattern shows 7 MHz inverted-V with apex at 45 feet and ends 10 feet clear of ground. Antenna wire runs in 90 degree azimuth plane. (Pattern from MN Antenna Analysis Program of K6STI.)

Chapter 4

Dipole and Ground Plane Antennas

(Two Popular Antennas and All About Them!)

The Dipole Antenna

The basic shortwave transmitting and receiving antenna is the *half-wave dipole* shown in Figure 1A. Its length is equal to about one-half the transmitting wavelength, and it is usually made of wire, supported at the ends by insulators. The dipole is broken at the center, at which point a two-conductor feedline is attached. The dipole can be mounted in any position, but it is usually mounted horizontally or vertically between two supports. The length of a wire dipole for any shortwave frequency may be computed from:

$$\text{Length (feet)} = \frac{468}{\text{Frequency (MHz)}}$$

For example, the length of a half-wave dipole at 21.2 MHz (near the mid-point of the 15 meter band) is:

Length = 468/21.2 = 22.1 feet, tip to tip

When erecting the dipole antenna, it is important to make sure that the two-conductor feedline does not interfere with proper antenna operation. To accomplish this, the line is led away at right angles to the antenna, as shown, when the antenna is in a horizontal position. When the dipole is mounted vertically, and it is desired to bring the feedline away below the antenna, a different scheme is used. The bottom of the

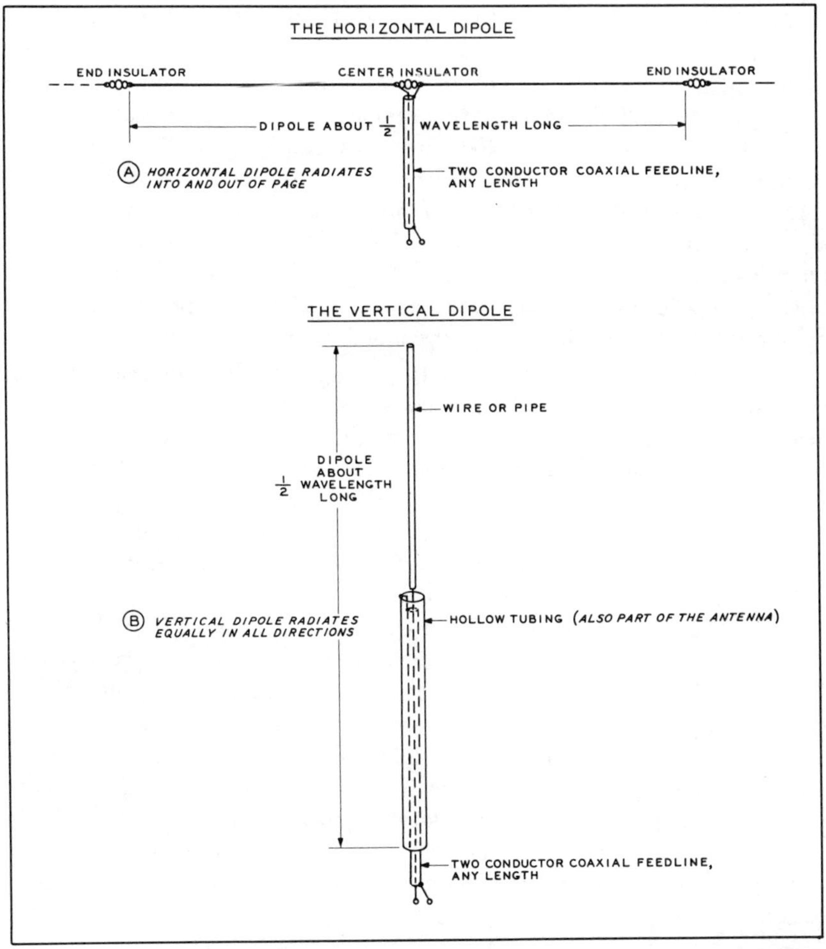

Fig. 1 THE DIPOLE ANTENNA is the fundamental shortwave antenna for both transmitting and receiving. Mounted horizontally (A), the dipole is commonly made of wire, supported at the ends by insulators. The low impedance center point is broken and a two conductor coaxial feedline is attached to the halves of the antenna. The feedline should be brought away at right angles to the antenna since it may have a radio field about it which can distort the radiation pattern of the antenna. This dipole radiates maximum energy into and out of the page.

The dipole antenna also may be mounted vertically (B). In order to bring the two conductor coaxial feedline away from the antenna without disturbing the radiation pattern of the antenna, the bottom half-section of the dipole is made of hollow tubing, with the insulated feedline running down the inside. Since the radio energy is concentrated on the outside of the dipole, the inside of the bottom section may be used as a shield for the feedline.

dipole section can be made of hollow metal tubing instead of wire, and the insulated feedline is passed down the center of the bottom section as shown in Figure 1B. Since the radio energy is concentrated on the *outside* surface of the dipole, the *inside* of one section may be used as a shield for the transmission line, provided the line does not touch the tubing wall. This effectively isolates the feedline from the electric field about the dipole. The dipole radiates radio energy at right angles to the direction of the conductor, as shown in the illustrations. The horizontal dipole thus has a bidirectional "figure-8" pattern and the vertical dipole has a nondirectional, circular pattern. As in many instances, the picture is worth a thousand words in this case.

Before we leave the dipole for the moment, it is important to note that this simple antenna is often considered to be the "standard" of comparison against which the performance of other antennas may be measured. There are other standards of comparison, and we'll discuss them when the time comes. But for now, remember the dipole as the comparison antenna for all practical purposes, and one of the most popular antennas for all-purpose high frequency communication.

The General Coverage Antenna – The Ground Plane

The simplest, low cost antenna for radio amateurs is the vertical, or *general coverage antenna*. This device radiates, or sprays, radio energy in all directions (except straight up). A typical general coverage antenna is the *ground plane*, shown in Figure 2. The ground plane consists of a vertical antenna, or *radiator*, mounted above several horizontal rods, or *radials*. The radiator and radials are all usually about a quarter-wavelength long. The length of a high frequency, quarter-wave element is:

$$\text{Length (feet)} = \frac{234}{\text{Frequency (Mhz)}}$$

The vertical section is considered to be the antenna proper, and the radials establish an artificial ground, or ground plane (hence the name) at the base of the radiator. Looking at it another way, the ground plane antenna is a vertical half-wave dipole, fed from below, with the bottom half of the dipole split into separate radials which are swung up into the horizontal plane. The antenna is fed with a two-conductor transmission line.

A general coverage vertical antenna allows you to communicate with other stations located in many different directions about you without the need of swinging a rotary beam back and forth.

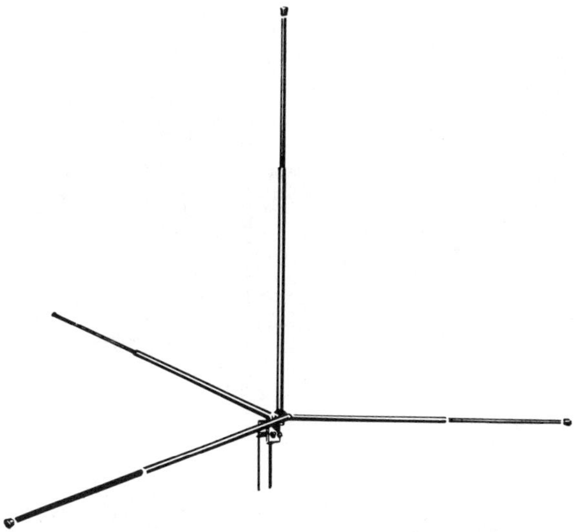

Fig. 2 THE POPULAR GROUNDPLANE. This inexpensive antenna is a close relative of the vertical dipole shown in Figure 1. It has a vertically polarized field and a nondirectional pattern, ''spraying'' radio energy in all directions about it (except straight up and down). The vertical rod is the antenna, and is a quarter-wavelength long. The three horizontal rods are termed ''radials'' and form an artificial ground (''ground plane'') directly beneath the antenna. At least two, and preferably three or four radials should be used.

The ground plane antenna is derived from the vertical dipole, with the bottom half of the dipole split into separate radials which are swung up into the horizontal plane. Some ground plane antennas have drooping radials which modify the electrical characteristics of the antenna so that it more closely matches a 50 ohm coaxial transmission line.

The radials may be connected directly to the supporting mast, as shown here, but the antenna portion of the ground plane must be insulated from radials and mast. The center conductor of the coaxial feedline is attached to the antenna, with the shield connected to the radials.

As mentioned earlier, the vertical antenna picks up a great deal more man-made interference than does a horizontal antenna. Automobile ignition interference, electric motor noise, power line "hash" and other radio "smog" are received more readily by a vertical than by a horizontal antenna. If you live near an industrial area and have power line interference--think twice before you put up a vertical antenna!

On the other hand, since a general coverage antenna receives and transmits in all directions, it does not provide the user with any protection from interfering signals on the frequency, since all signals are received regardless of the direction of the incoming signal.

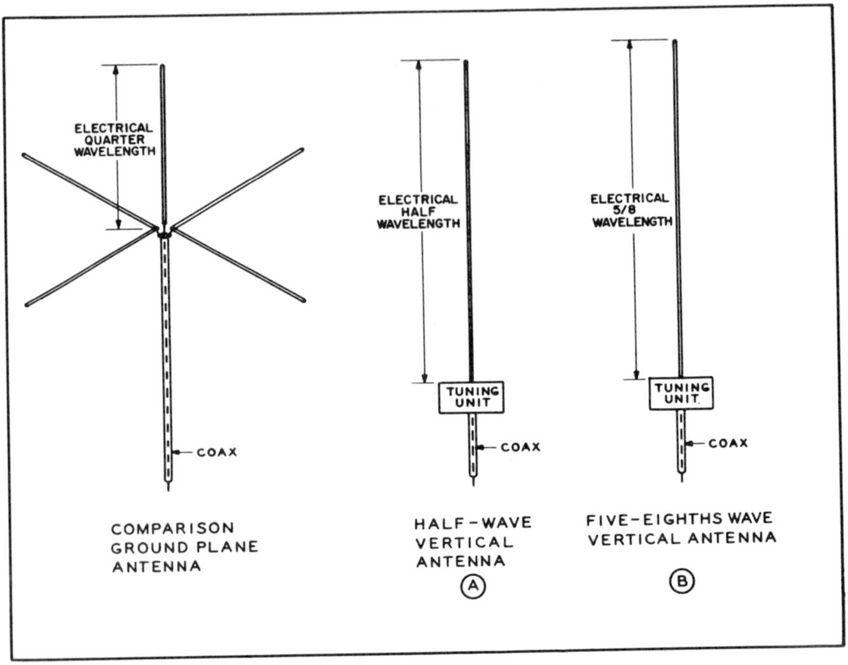

Fig. 4 EXTENDED VERTICAL ANTENNAS provide boost in gain over popular ground plane antenna (left). The half-wave vertical antenna (A) provides boost in power gain of 1.8 decibel over ground plane. The five-eighths wave vertical antenna (B) provides 3 decibel boost in power over a ground plane. This is a worthwhile increase in signal and an antenna of this type is recommended for general coverage work. No radials are required with the extended antennas. The main disadvantage of these antennas is that they are very responsive to automobile ignition noise and other man-made interference.

The Extended Vertical Antenna

Vertical antennas are not limited to one-quarter wavelength in length. In fact, the half-wave and five-eighths wavelength vertical antennas provide somewhat improved results over the more common ground plane. These antenna types are shown in Figure 4.

The half-wave vertical antenna is shown in illustration A. The dipole ends are points of high voltage and high impedance and the center point is at a relatively low voltage and low impedance. Some kind of matching device must be included to match this impedance to that of the transmission line. This device, or network, is mounted at the base of the antenna and may be included as part of the antenna itself, as described in a later chapter.

The half-wave vertical antenna is a complete dipole section and requires no radials in most installations. This is a great advantage where space is at a premium. The half-wave vertical antenna provides a power gain of about 1.8 decibel over the quarter-wave ground plane, a worthwhile increase in gain for such a simple antenna. The pattern is nondirectional, and general coverage is provided.

The five-eighths wavelength vertical is shown in illustration B. The extra length of the radiator provides additional power gain (a gain of 3.0 decibels over a quarter-wavelength ground plane) and also eliminates the need for the horizontal radials. The five-eighths wavelength vertical antenna is popular as a commercial broadcast antenna as it provides the greatest possible gain for a simple vertical antenna while still maintaining an omnidirectional pattern. It is also extensively used for industrial and police mobile service in the 27 MHz and VHF regions.

A five-eighths wavelength vertical antenna design for the 20, 15 or 10 meter bands will be described in a later chapter of this Handbook.

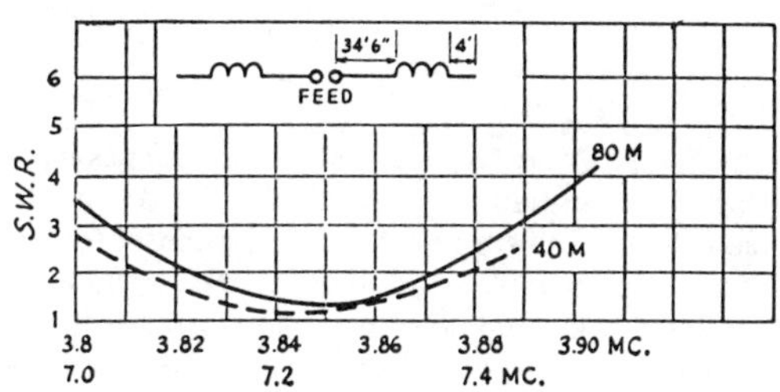

Fig. 5 80-40 METER DIPOLE uses rf chokes instead of tuned traps. Chokes are close-wound of #18 enamel wire on a phenolic form, 7/8-inch outside diameter and 14 inches long. Winding length is 12 inches. Inductance is approximately 120 μH. Resonance on 40 meters is determined by length of inner sections. Tip length determines 80 meter resonance. Idea was first proposed for amateur use by W4JRW. (Drawing from QST, April, 1961.)

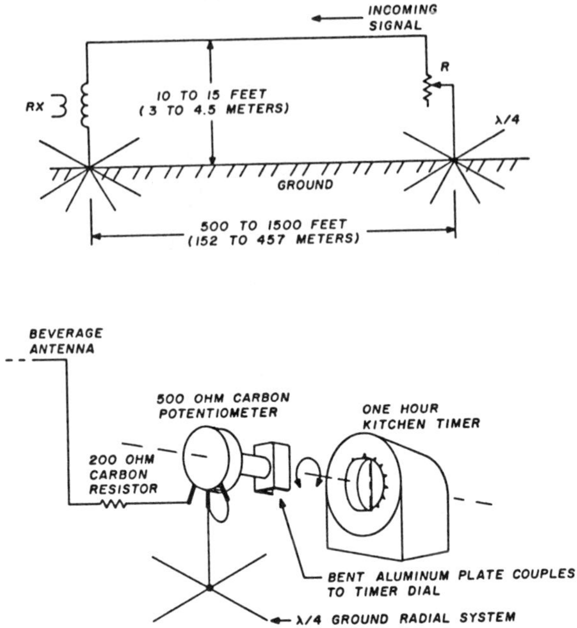

Fig. 5 K2GNC BEVERAGE ANTENNA for 160 meter reception provides directivity and cuts down on noise. Front-to-back ratio as high as 20 dB can be achieved. Value of resistor falls between 200 and 500 ohms. Potentiometer attached to kitchen timer is used. Receiver is tuned to broadcast station at high end of band located off back of the atnenna. Received signal is plotted against time to check resistance required for optimum front-to-back ratio as indicated by minimum S-meter reading. Potentiometer is then replaced with fixed resistor of proper value.

Chapter 5

The Coaxial Cable

(Your Radio Lifeline)

Your ham antenna installation is made up of two basic parts which should aid and assist each other: the *radiating system* (the antenna itself) and the *transmission line* (the feedline, a sort of "radio hose"). Each part has its own role to play and when each is designed to work properly with the other--and does--you'll have a successful and efficient antenna installation.

The antenna, as you know, is made of electrically conducting material, usually metal such as copper or aluminum. The mechanical supports for the antenna are made of material which does not conduct electricity, such as plastic, ceramic or wood. Materials of this class are termed *insulators*. Indeed, *if* the antenna elements are supported at the correct spot (the center of a dipole, for instance), the need for an insulated support no longer exists, no insulator is needed, and the element may be attached directly to a metal supporting structure. This is a great advantage in large beam antennas as a more sturdy and windproof array can be built if relatively fragile insulators are not needed.

The "Radio Hose"

At your station, radio energy from your transmitter is coupled into the feedline and carried by the line to the antenna. The feedline, therefore, may be thought of as a sort of "radio hose" through which radio signals flow efficiently from one place to another. A "lossy" or inefficient feedline, which offers resistance to the passage of radio energy, may be compared to a stopped-up garden hose, which slows down and chokes up the water before it arrives at your pet rose bush. The high resistance feedline, in addition, weakens (attenuates) the

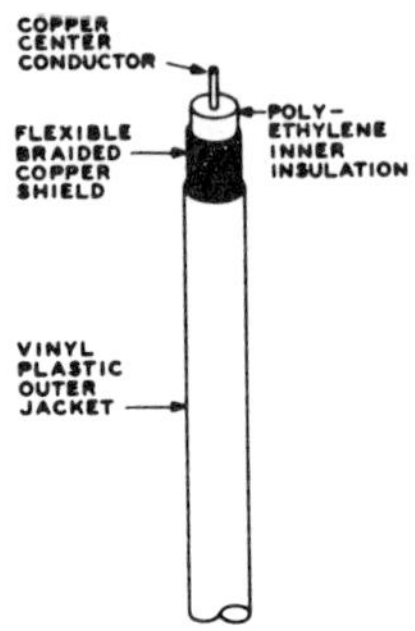

Fig. 1 COAXIAL CABLE is your "radio hose" through which radio energy flows to and from your antenna. The "coax" line consists of an inner conductor surrounded by a low loss insulating jacket and a braided copper outer conductor. The cable is covered by a waterproof vinyl jacket extruded over the braid. The inner conductor transmits the radio energy from one point to another and the outer conductor prevents the energy from escaping from the cable.

incoming signal before it arrives at your receiver. In short, a poor feedline severely cripples your amateur station, both for receiving and for transmitting!

The Coaxial Cable -- What Is It?

The *coaxial cable* (or concentric cable) is the most popular feedline used for radio transmission today.Over 500 various types of coaxial cable are manufactured, from microscopic, hair-like feedline for special instrumentation to giant line, several feet in diameter for super-power transmitters.

Coaxial line was originally designed at the turn of the century for use as submarine telegraph cables, and was modified for radio use in the early "thirties". Today's excellent, low-loss line was developed during World War II, and millions of miles of "coax" have been manufactured and used in the last 25 years.

An industry-wide standard for coaxial cable has been adopted, and most lines suitable for radio use are designated by the initials *RG* (derived from "radio guide"), followed by a serial number identifying a particular type of cable, and the suffix letter *U* (indicating "utility" cable). By far the most popular cables for amateur h-f or vhf operation are the RG-8/U and RG-58/U (50 ohm) families of cable.

The flexible coaxial line (Figure 1) consists of a solid or stranded copper inner conductor, a solid or perhaps white foamy continuous insulating jacket (usually polyethylene) and a braided copper outer conductor. The outer conductor is covered by a tough, waterproof vinyl jacket extruded over the braid. The purpose of the inner conductor is to transmit the radio energy from one point to another and the braided outer conductor (shield) prevents radio energy from escaping from the line. When properly used, no radio energy exists on the *outside* surface of the line which may, in fact, be buried beneath the earth without affecting antenna performance. The insulating jacket protects the coaxial line and prevents it from being accidentally shorted to ground or to other wiring.

The coaxial line is flexible enough to be bent around a radius greater than six inches or so. Sharp bends should be avoided, as the center conductor can "cold flow" or slowly move about within the insulation when under continuous stress, and may gradually shift over and short itself out to the braid.

The Black Jacket

No, the Black Jacket is not a Japanese Judo Society, but the outer insulation of the coaxial line. Two types of jackets are used, and you should steer clear of one. The jackets are *regular vinyl* and *noncontaminating vinyl*. Older cables (made up to about 1965, and a few inexpensive cables still made today) use regular vinyl jackets which seem to have a chemical affinity for the polyethylene inner insulation of the line. Result: after a period of four or five years of life (and sooner at high temperatures) the radio frequency loss of the cable increases markedly and it becomes almost useless as a "radio hose".

How do you tell "lossy" cable? You can't unless you are well equipped with laboratory instruments, or can conduct a simple measurement as described later in this Handbook. The recommendation is: Don't buy surplus coaxial line, or cheap line made by an unknown manufacturer. The military surplus line has been sold by Uncle Sam because its useful life has been exceeded. Lossy coaxial cable is a perfect way to throw your costly transmitter power away!

Coaxial Cable Loss

Cable loss is measured in decibels, and may be thought of as "negative gain". As an example, suppose you have an antenna with a power gain of 3 decibels, and you have a coaxial line running to the antenna that has a power loss of 5 decibels. Your overall power gain is negative, and you end up with 3 - 5 = minus 2 decibels power loss!

Cable loss is directly proportional to the length of cable you use, all other factors being equal. You should therefore strive to have as short a run of coaxial cable between your station equipment and the antenna as possible.

Coax Cable	Jacket	Dielectric	Nominal Impedance (in ohms)	Velocity Factor	dB Loss per 100 feet at 100 MHz	Diameter (in inches)
RG-8 (8237)	PVC	Solid	52	.66	2.2	.405
RG-8 Type (8214)	PVC	Foam	50	.78	1.8	.405
RG-8A (9251)	Long-life PVC	Solid	52	.66	2.2	.405
RG-213 (8267)	Long-life PVC	Solid	50	.66	2.2	.405
RG-8X (9258)	PVC	Foam	50	.78	3.7	.242
RG-58 (8240)	PVC	Solid	53.5	.66	4.5	.195
RG-58A (8259)	PVC	Solid	50	.66	4.9	.195
RG-58A Type (8219)	PVC	Foam	50	.78	4.5	.195
RG-58C (9203)	Long-life PVC	Solid	50	.66	4.9	.195

Fig. 2 COAXIAL CABLES recommended for ham use. RG-8 and RG-213 series are for high power or long cable runs. For short runs and ease of installation, smaller diameter RG-8X is recommended. The RG-58 family of coax is useful for power levels under 500 watts in the HF spectrum. Vinyl (PVC) outer jackets have a nominal life of five to ten years, limited by exposure to sunlight and gradual contamination of inner insulation. More expensive cables have "long-life" non-contaminating PVC jacket with an expected life of ten to fifteen years. Cable attentuation is progressively lower for lower frequency bands.

The Noncontaminating Jacket

In case you are buying coaxial cable, you should buy the newer type that has the noncontaminating jacket that does not affect cable life. The new jacket style designation for RG-8/U is RG-8A/U. An even newer cable, of the same type with a noncontaminating jacket is RG-213/U, which has the same characteristics as RG-8A/U. The new jacket style designation for RG-58/U cable is RG-58A/U, or RG-58C/U. These cables are all tabulated in the chart of Figure 2.

The newer coaxial cables with a noncontaminating vinyl jacket have a life expectancy in excess of fifteen years. Considering that the extra cost of the improved cable runs only about a penny a foot, the non-contaminating types are a good investment for your ham station.

Foam Dielectric Cable

A recent addition to the family of coaxial cables is the *foamed dielectric* type, designed for use with cable TV (CATV). The inner insulation consists of cellular polyethylene, foamed with an inert gas. While well-suited for their original purpose, their use at ham frequencies is really not worth the extra cost as the improvement in efficiency is only a percent or so over the less costly solid dielectric cables.

Your Choice of Cable

Figure 2 tells the story. Your ham equipment requires a 50 ohm transmission line. Actual coaxial line impedance runs from 50 ohms to nearly 54 ohms, but this manufacturing variation makes little difference in actual use. You are concerned with cable efficiency and long life. Vetoed are the cables with contaminating jackets, and those older cables with higher attenuation than others. The best group of cables to choose from are RG-8A/U and RG-213/U in the large diameter types and RG-58/U in the small diameter type. At 10 meters or lower in frequency, the efficiency of a 100 foot run of the RG-8A/U cable is about 80 percent, and the efficiency of the smaller cable is about 65 percent. The moral is obvious: keep your transmission line as short as possible, especially at 10 meters and the higher frequency bands of 6 & 2 meters.

Line Impedance

Most popular coaxial lines are designated as "50 ohms" or "70 ohms", or some such cryptic value. Often, this value is embossed into the jacket of the line. What does it mean? This reference does not refer to the actual resistance of the line in ohms (which is very low) but to the *characteristic impedance* of the line, which is an electrical description of the line in terms of the *ratio of line voltage to line current*. (As a refresher in this topic, better look once again at Chapter 3 of this Handbook). As before, the term impedance refers to a *ratio* and not to a numerical value of resistance. The characteristic impedance ratio is determined only by the physical characteristics of the line: the inner conductor diameter, type of insulating material, and the thickness and construction of the outer shield. By varying the ratio of the diameter of the inner conductor to the shield conductor and changing the inner insulation, lines of 50 ohm impedance to 90 ohm impedance may be manufactured.

The *resistance* of the line, on the other hand, is altogether a different matter. While complex laboratory instruments are required to measure the characteristic impedance, only an ohmmeter is needed to measure the line resistance. For a short section of good coaxial line, the resistance of either the center conductor or the shield (as measured from one end of the line to the other) will be a fraction of an ohm.

What is the Correct Line Length?

How long should your coaxial line be? Is there a "magic" length which will make your antenna work better? Generally speaking, the answer to this question is: no. The line should be long enough to reach from the station to the antenna, and no longer. The shorter the line, on the other hand, the less will be the line loss.

Two exceptions to this general statement should be noted: First, some ham gear exhibits erratic tuning when operated into a transmission line having a high value of SWR on it. The solution is to lower the SWR by making proper tuning adjustments to the antenna or to alter the length of the transmission line a foot or so so that the ratio of voltage to current at the transmitter end of the line is more in accord with the limitations imposed by the output network of the equipment.

Second, some beam antennas use short lengths of coaxial line as coupling devices between various portions of the antenna. In this case, the antenna manufacturer usually specifies a particular line length to be used. It is wise to follow his instructions. Aside from these two special cases, follow Abraham Lincoln's answer to the question, "How long should a man's legs be?" Honest Abe replied, "Long enough to reach the ground."

Is Coaxial Line Waterproof?

Yes! The line itself is waterproof as long as the tough outer vinyl jacket is not cut. *But the ends of the coaxial line are not waterproof* and--contrary to popular wisdom--*water can easily get inside your coaxial line at the exposed end.* If it does, you are in trouble! The water will be sucked along the outer copper braid by capillary action until finally the whole line becomes soaked inside. The copper braid gradually corrodes, ruining line efficiency and reducing the capability of your station. This capillary action is very nerve-wracking and unless you live in Phoenix, Arizona (where it rarely rains), the open end of your coaxial line can suck moisture out of the air over a period of time, just as a small child sucks a cherry soda through a straw!

Remember, then, when using any coaxial transmission line you *must* waterproof the ends of the line to prevent moisture from entering.

How To Waterproof Your Coaxial Cable

It's a cinch to keep your transmission line dry on the inside. Just don't let water get into it in the first place by properly sealing the ends of the line. The *wrong way* to terminate your line is to peel back the waterproof jacket and make a connection to your antenna using the inner conductor and shield as wire leads wrapped around the antenna terminals. "Shucks", says Johnny Knowitall, "Why should I spend 70 cents for a coaxial connector? I can just twist the wires around the antenna bolts, and it works great!"

Poor Johnny. Moisture will condense at the end of his line when the air temperature drops and water will gradually work its way down the outer braid of the line, inch by inch. Rain water enters the coaxial line directly. Soon, *alle ist kaput*.

(A note from the author: I had this happen to me once, when I lived near New York City. I'd had an unsealed coaxial line up for a few years and took it down one day to move the antenna. When I dropped the end of the line on the ground, water ran out of it. It hadn't rained for weeks, and I couldn't imagine where the water came from. Then I learned that coaxial line can act like a water pipe! I drained nearly a glass of water out of the line. When I cut the line open, the copper braid was corroded, and a nasty green color. The inner insulation was discolored too. That taught me a lesson--and an expensive one, as I had to replace all the lines from the house to the antenna.)

Well, to get back to the subject: If you have a coaxial plug on the end of your line it is necessary to coat the plug and the receptacle with a waterproofing silicone sealant, such as General Electric RTV-102. This flexible silicone rubber comes in a ready-to-use tube, and adheres to almost any surface. It can be bought in white or translucent shades (either is OK) and dries to a tough, flexible coating that resists moisture and rot. The silicone is applied over the coaxial plug and receptacle after connection is tightly made, and allowed to dry for a few hours. When still "sticky", it is overwrapped with black, vinyl tape, such as *Scotch plastic electrical tape No. 88*. Application of the silicone compound, plus a double layer of tape, carefully wrapped will do the job.

If your line is terminated in leads instead of a coaxial plug, the silicone sealant should be carefully spread over the base of the leads

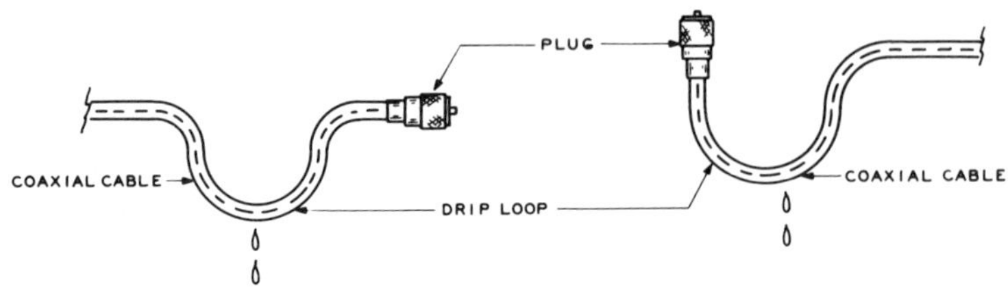

Fig. 3 COAXIAL CABLE IS WATERPROOF but most inexpensive fittings are not. In order to prevent water from entering the end of the line a drip loop should be used. In addition, the plug should be covered with a waterproofing compound and wrapped with vinyl tape, as discussed in the text.

and pushed down into the end of the cable. The end of the cable itself is then carefully wrapped with vinyl tape. Use two layers of wrapping.

The Drip Loop

A *drip loop* such as shown in the illustration (Figure 3) keeps the water from flowing down the line into the coaxial fitting. Always adjust the placement of the cable so that water tends to run away from, and not towards, the coaxial fitting. Allowing water to run into a coaxial plug is asking for trouble, no matter how carefully you prepare the joint. Finally, tighten the coaxial fitting to the receptacle with pliers to make sure the outer ring of the plug established a good connection between the shield of the line and the frame of the receptacle.

Never splice two coaxial lines together without the use of the proper fittings. Properly prepared lines with UHF fittings may be spliced together with a minimum of fuss with the use of the double-ended coaxial fittings listed in Figure 4.

The Coaxial Plug: Your Best Friend

A bewildering collection of coaxial plugs is manufactured for all types of coaxial cable. By far the most popular are the so-called "UHF fittings" listed in figure 4. Most modern ham gear uses these small connectors. They are not too expensive, moderately easy to install

COAXIAL CONNECTOR TABLE		
"UHF" CONNECTORS FOR RG-8/U, RG-8A/U AND RG-213/U LINES		
DESCRIPTION	MILITARY TYPE	AMPHENOL TYPE
PLUG	PL259 PL259A	83-1SP 83-1SPA
SOLDERLESS PLUG	— —	83-151
SPLICE	PL-258	83-1J
REDUCTION ADAPTER FOR RG-58/U	UG-175/U	83-185
"BNC" CONNECTORS FOR RG-58/U, RG-58A/U AND RG-58C/U LINES		
DESCRIPTION	MILITARY TYPE	AMPHENOL TYPE
PLUG	UG-88/U UG-88B/U UG-88C/U	31-002 31-018 31-202
SPLICE	UG-914/U	31-219
ADAPTER TO UHF	UG-273/U	31-028

Fig. 4 COAXIAL PLUG, splice and adapters make life easy for you and give you a ship-shape antenna system when they are properly used. This table lists both military and Amphenol type numbers of most popular coaxial hardware. Other types and styles of fittings exist and long lists of special coaxial plugs and adapters are given in the larger radio parts catalogs. Special connectors are also available to match UHF and BNC fittings to RCA and Motorola plugs.

(if you know what you are doing), but they are *not* waterproof. Figure 4 shows a listing of the most-used UHF and BNC fittings for both RG-8A/U (large) and RG-58A/U (small) 50 ohm cable.

No doubt many fingers have been burned and many swear words have defiled the atmosphere from the frustration of incorrectly installing coaxial connectors. You can put 'em on the cable the hard way, but here's the *easy way* to do it!

How to Use Coaxial Plugs -- The Easy Way

Make neat and efficient connections between your coaxial cable and your equipment! Many radio amateurs look upon coaxial plugs as inventions of the devil, and they may be right. Tricky to place on the line, when improperly installed, the common coaxial plug can cause intermittent operation and possible damage to the transmitter portion of your ham equipment.

Getting the coaxial plug properly fitted on the end of a coaxial cable may be a frustrating and time consuming task, in spite of all the instructions and drawings of a correct assembly shown in magazines and handbooks. These wishful directions usually don't work. In many cases, the ham simply gives up, jams the connector on the cable, leaving short whiskers of copper braid ready to short out the antenna circuit -- and an open invitation for failure!

Properly done, the job of placing a coaxial plug on the transmission line is not difficult. The following assembly sequence was worked out over the years and is recommended for the popular type PL-259 coaxial plug, obtainable at any large radio supply house. This connector is intended for use with RG-8/U cable and (with the addition of an adapter) also with the smaller RG-58/U cable. You can determine if this style of plug fits your radio by looking at the antenna jack. If it is labelled SO-239 or 83-IR, it will match the PL-259.

The first step in preparing the coaxial cable for the plug is to slide the coupling ring of the plug onto the cable with the ring threads towards the open end of the cable. Next, take a utility knife (*Stanley 99A Shop Knife*, for example) and circumscribe a cut in the outer black jacket of the line. Make the cut at right angles to the cable and about 1½ inches back from the end of the line. The small cylinder of jacket material you have cut may be slit carefully with the knife and removed.

You have now exposed over an inch of the outer copper braid of the cable. Without disturbing the braid, which should be lying flat against the inner insulation, take a soldering gun and quickly and smoothly tin the exposed braid, making it a solid entity. Don't overheat the braid, or the inner insulation may melt and "squirt" out between the basket-weave strands of the braid. One or two practice runs on scrap cable ends will make an expert of you! Clean the left-over flux from the braid with a rag moistened with paint thinner or alcohol (See Figure 5).

The next step is to trim the soldered braid to the correct length. Use a small tubing cutter for this job. The *General Hardware #123 Midget Tubing Cutter* is recommended. Cut the braid so that 7/16 inch is left on the cable end. Mark a line this distance from the black vinyl jacket and place the tubing cutter over the braid, letting the cutting wheel fall on the mark. Tighten the cutter slightly and slowly revolve it about the cable. After one turn, tighten the wheel again, and continue to revolve the cutter. Four or five turns and the cutter will neatly slice the soldered braid. The unwanted slug of braid may be then snipped off with a pair of wire cutters. In using the cutter, don't cut down too far, or you'll slice into the center insulating material.

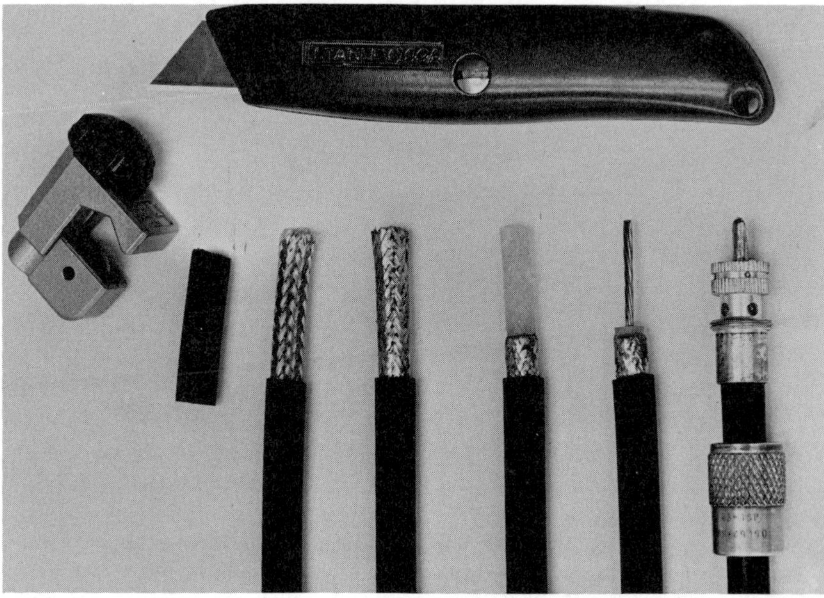

Fig. 5 COAX PLUG ONTO LINE--THE EASY WAY. This photograph shows the easy steps in preparing RG-8A/U coaxial line for a PL-259 coaxial plug. Midget tubing cutter (left) and utility knife (top) are used. At left is sample coaxial line with the outer jacket removed by the knife. Next, the outer, braided conductor of the line is tinned. Third view shows the outer braid cut to proper length by the tubing cutter. Fourth view shows the inner insulation of the line cut to length and inner conductor tinned. Right view shows coaxial plug and ring on line, with plug in position for soldering to line. Soldering is done through four holes in shell of plug. Soldering gun or iron with small tip and high wattage rating is recommended for this operation. Once the plug is soldered, it may be wiped with damp cloth to remove resin and to cool it.

The next step is to trim the center insulation. Cut it cleanly with the utility knife so that a collar 1/16 inch wide extends beyond the soldered braid. Easy! Don't nick the center conductor. Once the insulation is cut, you can pull the slug off the end of the cable by grasping it with your fingers and gently pulling it, rotating it at the same time so that it follows the twist of the inner conductor wires. When the slug is off, tin the center conductor.

Now the cable end is ready for the PL-259 plug. Push it carefully on the cable end, rotating it with the fingers so that the internal threads of the plug screw onto the outer vinyl jacket of the cable. Make sure the inner conductor is centered into the plug pin. As the plug body is

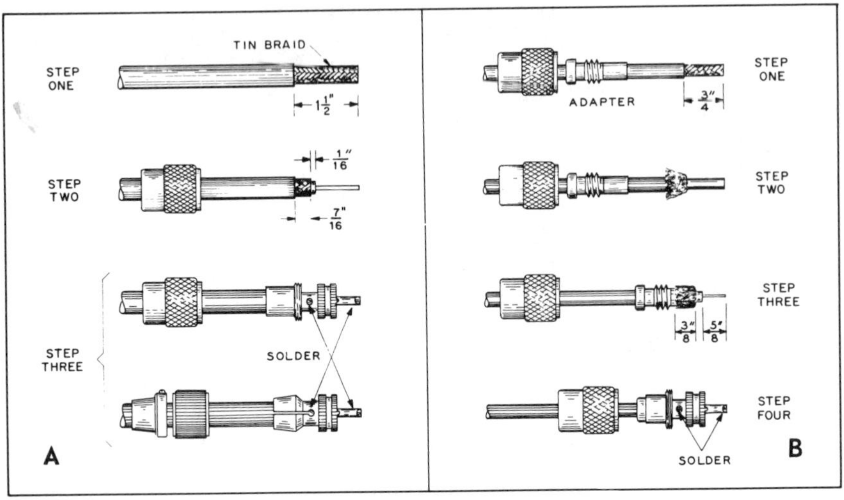

Fig. 6 UHF PLUG AND ADAPTER installed on a coaxial line -- the easy way. PL-259A plug installation is covered in (A), corresponding to the steps shown in Fig. 5. PL-259 and UG-175/U adapter are shown in (B) matching the small diameter RG-58A/U coaxial line to the large diameter coaxial plug.

screwed onto the cable, you'll see the tinned braid appear through the four solder holes in the shell. Continue twisting the plug onto the cable until the braid is completely visible through all holes (Figure 6A).

Now is the time to solder the plug onto the cable, the idea being to solder the braid through the four solder holes. Use a soldering gun with a small tip and proceed with care, using small diameter solder. It is usually easier to hold the plug in a bench vise during this operation. Take care that solder does not run over the outer threads of the body. The last step is to solder the center conductor to the plug pin. After the assembly cools down, slide the coupling ring down over the plug.

Try this technique on a plug and a spare piece of coaxial line. Soon you'll be able to produce masterpieces that will arouse the envy of your friends. A real artist, once he masters this simple technique, can then advance to the reverse step -- salvaging coaxial plugs from old cable!

The PL-259 Plug with Small Coaxial Cable

The PL-259 may be used with small diameter cable (RG-58/U, for example) by adding a reduction adapter (type UG-175/U). A slightly different assembly technique is used for this cable (Figure 6B).

The end of the cable is passed through the coupling ring and the adapter, with the threads of the ring and the narrow end of the adapter facing the end of the cable. Using the utility knife, cut 3/4 inch of the vinyl jacket off the cable. Fan the braid out slightly and carefully fold it back over the adapter.

Next, trim the braid with a small scissors to about 3/8 inch long, so that it fits about the barrel of the adapter. Following this, take the utility knife and remove 5/8 inch of the insulation from the center conductor. Careful! Don't nick the conductor. Finally, tin the conductor.

Now, carefully screw the plug body onto the adapter. The center conductor of the cable should pass easily through the center pin of the plug, and the strands of the braid should appear through the side holes of the shell. Using a small soldering gun, solder the braid through the holes. Lastly, solder the center conductor to the plug pin and slide the coupling ring down over the plug.

Check Your Coaxial Line!

The first step to take when antenna trouble is suspected is to carefully check your coaxial line, especially the plugs at both ends. By far the greatest number of antenna difficulties stem from incorrectly soldered coaxial line fittings. Then, too, in checking an old feedline installation, or checking someone else's line, the end fittings should be carefully examined.

You can perform an important check on your line with the aid of an inexpensive volt-ohmmeter (Figure 7). The ohmmeter should be set on the *highest resistance* range and the coaxial line disconnected at both ends. By measuring the resistance between outer braid and inner conductor at one end of the line you can make sure that a leakage path does not exist *across* the line from braid to inner conductor. The resistance reading between inner and outer conductor on a good piece of coaxial cable should be *infinity.* If a resistance value of a few hundred ohms (or several thousand ohms, for that matter) is noted, the coaxial fittings should be immediately suspect. In addition, the line itself should be closely examined for abrasions, breaks in the jacket, and damage to the copper braid under the jacket. If a coaxial plug is in doubt, it should be cut off, discarded, and a new one substituted in its place.

Finally, using the ohmmeter on the *lowest resistance* range, measure the cable from one end to the other along the same conductor. Both the outer braid and inner conductor should show an end-to-end resistance of less than one ohm or so. Thus, a good coaxial line shows infinite re-

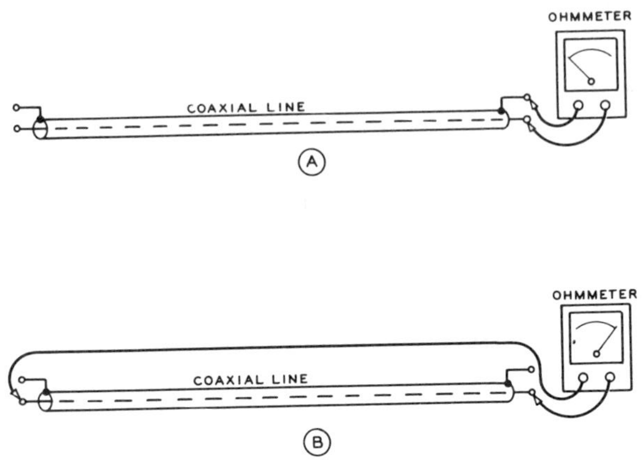

Fig. 7 **TEST YOUR COAXIAL LINE** with an ohmmeter. Leakage across line is measured between conductors (A) with meter on highest scale. Resistance of each conductor is measured along conductor (B) with meter on lowest scale.

sistance *between* the conductors and near-zero resistance *along* each conductor from end to end.

The final coaxial line check is to run transmitter power through the line into a *dummy load* and see how things perform. Place the dummy load at the far end of the line, hook the opposite end of the line to your equipment through an *SWR meter*. Tune up the equipment in the normal manner. A good coaxial line will show practically no reverse reading on the SWR meter after the instrument is set for a forward, full-scale reading (in other words, the SWR is unity, or 1). If the coaxial line passes this test it is A-Okay and will deliver power to your antenna. There's more on the handy SWR meter in the next chapter.

Line Loss Vs. SWR

The line attenuation given in Figure 2 assumes that the SWR on the transmission line is unity. What happens to line loss at high values of

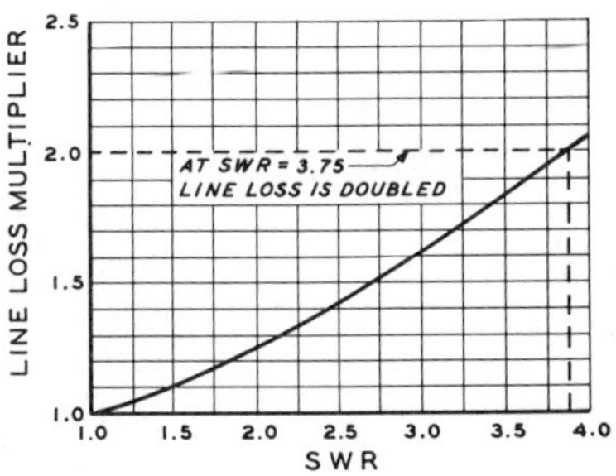

Fig. 8 COAXIAL LINE LOSS increases as SWR increases. Line loss will be doubled when SWR is 3.7. Reducing SWR drops line loss. Lowest loss is achieved by using large diameter coaxial line such as RG-8A/U.

SWR? The answer is that line loss increases with increasing values of SWR as shown in Figure 8. For example, if a transmission line has an SWR value of 3.7, the line loss will be doubled. A 100 foot length of RG-58A/U, for example, has a nominal line loss of 1.9 decibel. If the SWR is 3.7, the line loss increases to 3.8 decibels. This corresponds to a 60 percent loss of power! Changing from RG-58A/U to RG-8A/U will drop the line loss from 3.7 decibels down to 2 decibels, which corresponds to a 38 percent loss of power.

For long runs of transmission line, it is prudent to use the heavier and more expensive RG-8A/U line rather than the smaller RG-58A/U cable. In any event, the overall length of the transmission line should be held as short as possible.

The above attenuation figures are for 10 meters. The figures are lower for the low frequency bands, and quite a bit higher for the 6 and 2 meter bands. The small diameter RG-58A/U line, in fact, should not be used on 2 meters for runs of much greater than 10 feet or so because of excessive loss. Runs of RG-8A/U up to 50 feet or so are permissible at 2 meters without undue loss of power. At all frequencies, line loss increases with SWR as shown in Figure 8.

Power Capability

All types of coaxial transmission lines are power limited, with the power capability decreasing as the frequency increases. In the amateur high frequency bands RG-58 and RG-59 families of cable may be safely used with transceivers and other equipments up to the 500 watt power level. At powers in excess of this, the heavier RG-8 and RG-11 families of cables should be used.

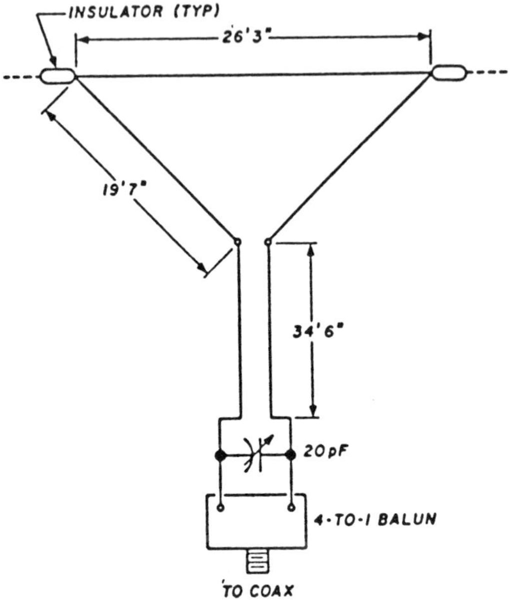

Fig. 9 DELTA LOOP for 7 through 28 MHz. This antenna, designed by HB9ADQ, covers 7 amateur bands. The 20 pF capacitor across the balun is adjusted for lowest value of SWR. On 18 and 24 MHz the balun must be replaced with a balanced antenna tuner.

The feedline consists of #16 wire, spaced 4 inches by means of spreaders cut from lucite rod. The antenna is hung in the vertical plane and the radiation pattern is a figure-8, at right angles to the plane of the loop.

Chapter 6

The SWR Meter

(Friend or Foe?)

How can you be sure your antenna is operating at its highest degree of efficiency? Unlucky Pierre, the well-known DX operator says, "Yes! My antenna is perfect, but I cannot talk to anyone."

What's wrong with Unlucky Pierre's antenna? Is he merely the victim of bad luck and interference on the frequency, or he has made a monumental "goof" in assembling his antenna that has rendered his expensive station useless?

In a few moments Pierre can get an answer to his antenna problems--which may be real, or merely imaginary--by making simple measurements with an instrument known variously as an *SWR meter, reflectometer,* or *SWR bridge* (Figure 1). This device (placed in the transmission line) reads the *standing wave ratio (SWR)* of the antenna system on a meter and by measurement of this ratio tells Pierre how well his antenna system is working. It will tell you the same thing, too!

Incident And Reflected Waves

Radio waves travelling through space, or along a transmission line, have been compared to water waves made by a stone thrown into a quiet pond. The waves travel outward from the stone in expanding, concentric circles until they meet an obstacle, at which point they are reflected back towards the source. The forward travelling water waves *(incident waves)* and the *reflected waves* interact with each other and form interesting patterns of interference on the surface of the water.

Another fanciful analogy to radio waves is of a person holding a long rope attached to a post. He can start flipping the end of the rope, creating a wave of motion moving along the rope towards the post. Upon reaching the post, the wave will be reflected back towards the opposite

Fig. 1 THE SWR METER is useful instrument for your ham station. It measures the ratio of the incident (forward) to the reflected power in your transmission line. This ratio is termed the standing wave ratio. The SWR meter is placed in series with your transmission line between the ham transmitter and the antenna. Power is applied from the transmitter and the potentiometer control is set for a full scale meter reading when the control switch is in the "forward power" position. Switch is then thrown to "reverse power" position and the standing wave ratio is read directly from the meter of the instrument.

end. If a number of equally timed flips are given, a succession of waves at equal intervals are sent along the rope. When reflected back from the post at the far end, they meet others coming along whose "wavelength" is equal to that of the waves coming back. At some points the rope tends to move a certain distance upwards with the direct wave, and the same distance in the downwards direction with the reflected wave; the result is that at these points the rope does not seem to move at all. These points are found along the rope one half wavelength apart. At all other points the rope moves (or vibrates) in the resultant direction of the combined direct and reflected wave impulses, creating what are called *stationary waves*, or *standing waves*. Waves of this type can be set up along a conductor by suitably timed electrical impulses applied to the conductor. A picture of the standing wave analogy is shown in Figure 4, Chapter 2.

What Is The Standing Wave Ratio (SWR)?

A combination of incident and reflected electric waves on a radio conductor will result in a combined wave called a *standing wave*. The ratio of size (amplitude) of the incident to the reflected wave is termed the *standing wave ratio* (abbreviated SWR). The standing wave ratio may be measured by a suitable instrument.

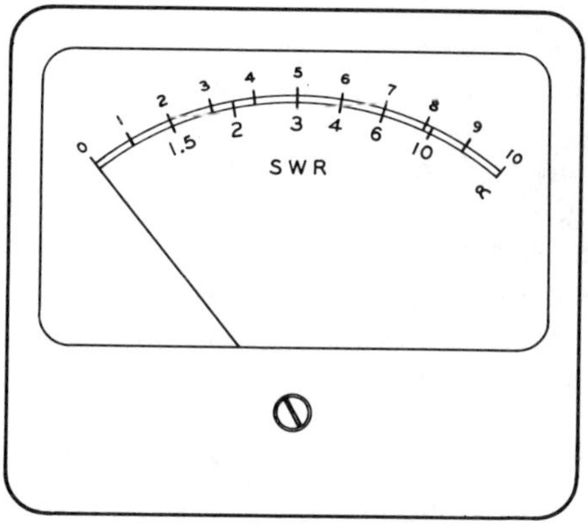

Fig. 2 TYPICAL SWR METER CALIBRATION. The top scale is merely for reference. Bottom scale is calibrated in SWR. Lowest SWR reading is at left of scale. SWR readings above 10 are usually not accurate and most meters are calibrated only to that value. SWR meter measures forward and reverse power in transmission line, but scale is calibrated in terms of SWR for easier interpretation.

The SWR on a coaxial transmission line is measured by an SWR meter and the magnitude of the SWR is read directly from the scale of the instrument. The SWR is expressed as a ratio, with unity (zero reading) indicating no reflected power and infinity (full scale reading) indicating a state of maximum reflection. A typical SWR meter scale is calibrated as shown in Figure 2.

Since the electrical impulses consist of voltage and current waves which follow each other regularly, the standing wave as a result, is composed of waves of voltage and current. As a definite relationship exists between the voltage and current waves, the amplitude, or strength, of the standing wave may be ascertained by measuring either the voltage or the current in the proper manner. Most inexpensive SWR meters monitor the voltages in a transmission line and compare the incident voltage with the reflected voltage.

With a perfect antenna system, the reflected wave is zero and SWR is termed to be *one, unity,* or *one-to-one*. As the reflected wave increases in amplitude, the SWR increases. The SWR reading, in effect, shows the *degree of mismatch* between the impedance of the transmission line and that of the antenna itself, as discussed in the following section.

Antenna Matching

When the radiation resistance of the antenna (that realistic voltage to current ratio) is the same value as the characteristic impedance of the transmission line (the ratio of voltage to current in the line), all of the electric energy transmitted down the line is taken by the antenna and radiated into space in the form of a radio wave. If the antenna does not match the transmission line perfectly, wave interference is set up at the junction of the line and the antenna and a portion of the energy is reflected by this mismatch point back down the transmission line towards the transmitter. As the degree of mismatch between the antenna and the line increases, the amount of reflected radio energy increases, just as a larger obstacle tossed in the water pond will reflect back a bigger wave of water. Standing waves on the transmission line are thus created by the interaction of the forward and reverse travelling electric waves.

As the reflected wave increases, the incident wave decreases, but the total power in the two waves remains the same. The worst possible case exists when the incident and reflected waves are equal in strength. This happens when the far end of the transmission line (the end away from the transmitter) is either short circuited or open.

The SWR Meter--How Does It Work?

With the idea of incident and reflected waves under our belt, it's not too difficult to understand the operation of the SWR meter, a schematic of which is shown in Figure 3. This version of the SWR meter (and there are others) takes advantage of the fact that the voltage and current of the forward wave are in step (in phase) while the voltage and current of the reflected wave are out of step (out of phase).

The SWR meter is placed in the coaxial line. It has two small pickup wires placed near the inner conductor of a section of line. Current is coupled into a pickup wire when a radio wave passes down the line and a very small amount of voltage is sampled through the capacity of the wire to the line.

The current in the pickup wire flows through a resistor at the end of the wire and the voltage across the resistor is measured by a sensitive meter. Two pickup wires are incorporated in the SWR meter, reversed with respect to each other. When the SWR on the line is low, the pickup wire voltages and currents are in step and are low in value. As the SWR increases on the line, the pickup wire voltages and currents are not in step, and are higher in value. The voltmeter consequently reads a resultant voltage. Thus, by using two pickup lines, reversed with respect to each other, currents and voltages flowing in either direction in the line may be compared. Since the SWR meter is symmetrical, either pickup unit may be used for either forward or reflected measurements and both readings may be applied to one meter via a selector switch.

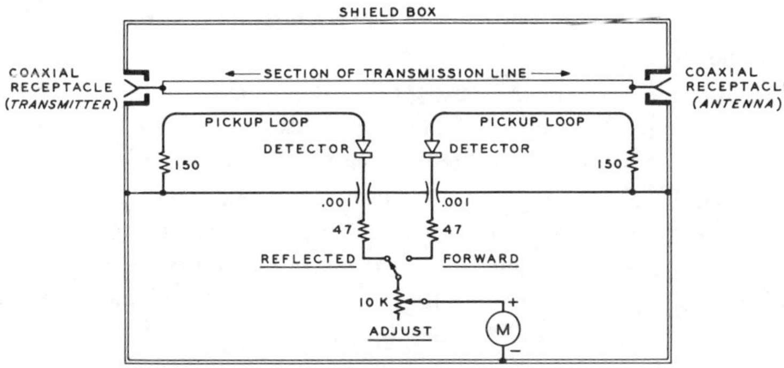

Fig. 3 SCHEMATIC OF TYPICAL SWR METER. Many different types of SWR meters exist, but this version is commonly used in inexpensive designs. The sensing device is a short section of transmission line inclosed in a shielded box. Twin pickup loops sense the voltage and current flowing in the incident and reflected waves. Selector switch permits operator to observe relative voltages on meter of instrument.

Using The SWR Meter

The comings and goings of the radio power in your transmission line can be neatly separated by the SWR meter and the power travelling in each direction can be read on the meter of the instrument. Adjustments may be made to the antenna in order to reduce the reverse (or reflected) meter reading, thus dropping the value of the SWR on the transmission line. *The name of the game is to reduce the reverse reading to as low a value as possible,* say, less than 10 percent of the forward reading, or to as near-zero as possible. Making such antenna adjustments without the use of an SWR meter is like washing your feet with your socks on-- you can do it, but it ain't easy!

The SWR Meter As A Tuning Aid

Why bother with an SWR meter? Well, the SWR meter is a powerful tool to aid you in obtaining optimum antenna performance. At the same time it may be used to tune your transmitter for maximum power output. Through maladjustment, it is possible to drop transmitter power output by as much as 90 percent, without the operator being aware of the loss of power! The proper use of the SWR meter will eliminate catastrophies of this type.

In addition to giving the operator a tool to determine proper transmitter tuning, the SWR meter provides a handy means of judging antenna adjustment. If the SWR on the transmission line to the antenna is high, it indicates a bad degree of mismatch between the antenna and the transmission line. *No adjustment or tuning done at the station can cure this difficulty,* only adjustments made to the antenna can possibly lower the SWR on the transmission line. Incorrect antenna adjustment causes a high value of SWR and this, in turn, causes an increase in power lost in the line and a drop in efficiency of your transmitter. Indeed, some "solid state" (completely transistorized) VHF FM transmitters will not operate into a transmission line having a high value of SWR. In most cases, the equipment has a SWR protection circuit that turns the gear off before damage is done to the transistor stages.

A Trial Run With Your SWR Meter

Let's make an actual SWR measurement with an inexpensive SWR meter, such as shown in the photograph. The SWR meter should be placed in series with your transmission line, near your operating position, so that a check may be made on both transmitter operation and SWR meter readings. Most inexpensive SWR meters are equipped with SO-239 type coaxial receptacles which match the common PL-259 type coaxial plug. Connect the output (antenna) receptacle of the SWR meter to your transmission line and antenna, and connect the input (transmitter) receptacle of the SWR meter to your set through a short, 50 ohm coaxial line about two feet long or so. One end of this line should have a PL-259 matching plug on it, and a plug suitable for your radio is placed on the opposite end. Since you, good reader, have just finished reading the section of this Handbook on the fine art of placing plugs on coaxial lines, we'll assume that your coaxial connections are properly made! Your complete installation is pictured in Figure 4.

Make sure that the coaxial line used between SWR meter and transmitter is the same type as the line running to the antenna. By far the greatest number of ham gear and antennas are designed for a 50 ohm line, and most SWR meters are designed for a 50 ohm line, too. However, 70 ohm line does exist, and some SWR meters are made for 70 ohm line. Don't mix up a 70 ohm SWR meter with a 50 ohm line, or you'll get funny readings that don't mean a thing! All bets are off, too, if you

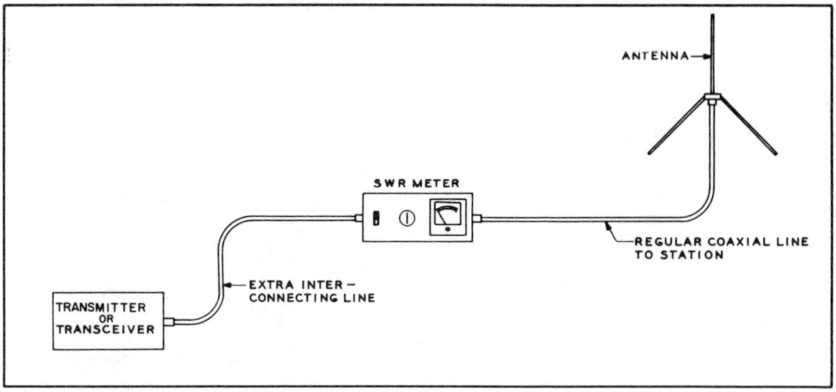

Fig. 4 TYPICAL SWR METER INSTALLATION. The instrument is placed in series with the coaxial line to the antenna, near the operating position. A short, extra line runs from the equipment to the SWR meter. The instrument may be left permanently in the line as a check on antenna and equipment operation. Sensitivity control on SWR meter permits use with transmitter powers up to maximum amateur limit. Use carrier injection for SSB tuneup.

try and make-do without the use of coaxial connectors, as you'll get screwy readings when the radio energy runs all over the place. Keep the energy *inside* the line where it belongs!

Once the connections to the SWR meter are properly made, the panel transfer switch of the device is set for a *forward* (FWD) reading, and the *sensitivity* control is set to minimum sensitivity (maximum counter-clockwise position). Turn on your transmitter and tune it up in the normal manner. Now advance the sensitivity control on the SWR meter for full scale deflection with the transfer switch set in the *forward* position. Without touching the sensitivity control setting, now throw the switch to the *reverse* position and note the meter reading. If all is well, the reverse reading will be substantially less than the forward reading. Most SWR meters have a scale calibrated in "Standing Wave Ratio" and the reverse reading is taken from this scale. If all is well with your antenna system the reverse reading will be less than 2. A really well adjusted (or "matched") antenna will present a reverse reading of less than 1.5 or so. A perfect match will result in a reverse reading of zero.

If your reverse reading is over 2, or--heaven forbid!--is nearly as high as the forward reading, you are in trouble! Either the instrument is incorrectly calibrated, incorrectly placed in the circuit, or the antenna is malfunctioning. A very high reverse reading usually indicates that either an open connection or a short circuit exists in the antenna system somewhere.

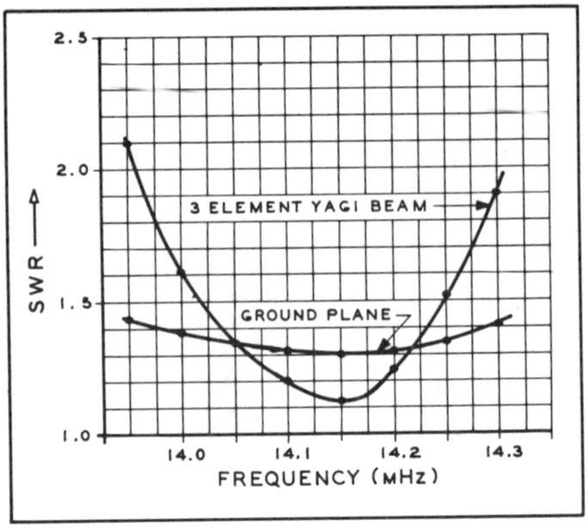

Fig. 5 TYPICAL SWR CURVES for Yagi beam and ground plane antenna. Chart is made up by noting SWR reading for a particular ham band, and plotting SWR versus frequency across the band.

You'll notice that when you run this test, the reverse SWR meter reading varies from frequency to frequency. This is normal, and you can actually plot a curve of SWR versus frequency for a ham band, as shown in Figure 5. The curve should be smooth and the minimum value of SWR should fall about the middle of the amateur band. You can make a separate SWR graph for each amateur band, logging both forward and reverse readings. Plot the reverse (SWR) reading on a sheet of graph paper and you'll have a record of the performance of your antenna. Later, if you suspect antenna trouble, you can re-run the SWR measurements and compare them with the original curve.

Transmitter Tuning With The SWR Meter

The SWR meter may be used to peak the transmitter controls for maximum power output, or to check output. Merely tune your transmitter output stage for greatest meter reading with the SWR meter switch set in the *forward* position. The meter reading is not calibrated in watts, but merely gives greatest indication when the transmitter output is maximum. Make sure you do not drive the meter off-scale (retard the sensitivity control if necessary) or the instrument may be damaged.

Friend or Foe?

To repeat: the lower the reverse reading on the SWR meter as compared against the forward reading, the lower the SWR value of your antenna system. If the antenna match is perfect, the reverse reading will be zero. Most properly designed amateur antennas exhibit a reverse SWR reading of 2 or less at any frequency in the operating range.

The question might well be raised: just what is the maximum value of reverse SWR reading that may be encountered in a typical antenna installation while still assuming that the antenna is working properly? Generally speaking, if the SWR reading is greater than 1.5 or so, it is an indication of improper operation and (other things being equal) possibly could indicate that things are amiss at the antenna end of the coaxial line. A SWR of 2 or more is a definite indication of antenna malfunction. Good ham antennas hold the SWR across a ham band to an indication of about 2 or less, and SWR readings of this magnitude, or less, should be no cause for concern as the mismatch is considered trivial and may be ignored.

An exception to this statement is the short mobile mini-whip antenna used on the lower frequency ham bands with a loading coil. The mini-whip characteristics are such as to reveal a high value of SWR on the transmission line when the whip is operated at a frequency removed from that to which it is tuned. SWR values of 3 or more may be observed with a mini-whip antenna operated off-frequency as is often done with 80 meter mobile antennas.

Finally, we come face-to-face with the question at the front of this chapter: SWR Meter--friend or foe? The answer is that the SWR meter can be the best friend you ever had *if* you use it properly and can be your worst enemy if you don't understand the significance of the readings. Take time to measure the SWR of your antenna system on all the frequencies you use. Make a graph of the readings, as shown in the illustration. Keep the graph and re-check your readings every month or so. If trouble develops in your antenna, you'll notice that the SWR readings have changed. You can continuously check antenna operation by leaving the SWR meter in the transmission line. It will indicate your transmitter output power, and a glance at the meter once in a while is a great satisfaction as you realize your equipment is operating properly.

In a later chapter the use of the SWR meter in conjunction with tuned transmission lines and antenna tuners will be discussed.

Chapter 7

DX Dipole Antennas You Can Build

(Simple and Effective Antennas That Work!)

The horizontal, single wire antenna is one of the most popular forms of transmitting and receiving antennas. It may be either end-fed or center-fed. When it is center-fed and half wavelength long, it is called a *dipole*.

The dipole is a balanced antenna. That is, it is symmetrical about the center feed point (see Figure 1, Chapter 4) and current and voltage in one half of the dipole match those values in the opposite half. Both feedpoint terminals at the dipole center are electrically balanced to each other and with respect to ground.

Coaxial line is almost universally used to feed the dipole antenna. This line, however, is an unbalanced conductor, since the outer conductor is (or should be) at ground potential, as discussed in Chapter 5. For best antenna performance, a balun transformer is often connected between the line and the dipole to preserve antenna symmetry with respect to ground. This is desirable from a TVI (television interference) reduction standpoint and to avoid difficulties with unbalanced currents on the line, and consequent unwanted radiation from the transmission line itself. Construction information for a suitable balun is included in a later chapter in the Handbook, and such a balun may be added to the dipole antennas described in this chapter.

Connecting a dipole antenna directly to the coaxial line is frowned upon by purists, but it is done every day, apparently without drastic effects or loss of antenna efficiency. Unbalanced currents in the antenna and line apparently cause no great mismatch, even though antenna currents may flow on the outside of the line. Inclusion of a balun in these antenna designs, then, is left to the option of the builder.

THE DIPOLE ANTENNA

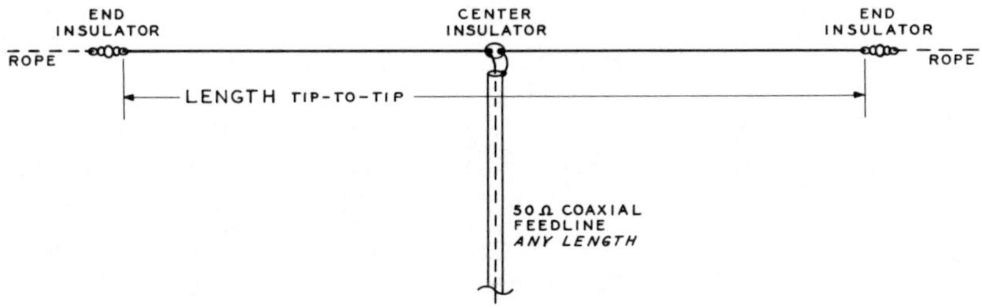

BAND	CENTER FREQ. (MHz)	LENGTH TIP-TO-TIP	METRIC LENGTH
160	1.85	253′ 0″	77.10
160	1.95	240′ 0″	73.15
80	3.60	130′ 0″	39.62
80	3.80	123′ 2″	37.54
40	7.15	65′ 6″	19.98
30	10.12	46′ 3″	14.10
20	14.15	33′ 1″	10.10
17	18.12	25′ 10″	7.87
15	21.20	22′ 1″	6.73
12	24.93	18′ 9″	5.72
10	28.60	16′ 5″	4.98
6	50.20	9′ 2″	2.80

$$\text{LENGTH (FT)} = \frac{468}{F\ (\text{MHz})}$$

Fig. 1 **DIPOLE ANTENNA FOR HIGH FREQUENCY BANDS** is tuned to frequency by adjusting tip-to-tip length of wire. Response is broad enough so that when antenna is cut for mid-band operation on 6, 10, 15, 20 or 40 meters it will function properly across entire band. Separate dimensions are necessary for low and high frequency portions of the 80 and 160 meter bands.

In the case of a high gain beam antenna, the inclusion of a balun between the balanced antenna element and the coaxial line is essential to preserve the electrical characteristics of the directional beam pattern. Front-to-back ratio and power gain of the beam both suffer if the feedline is not properly connected to the beam.

Since the dipole antenna has neither power gain nor front-to-back ratio, it may be directly fed with a coaxial line and will provide good results.

Used by thousands of amateurs the world over, the single band dipole is efficient, inexpensive and relatively unobtrusive. The dipole works on any one amateur band and is tuned to that band by adjusting the length of the wire flat top section, as tabulated in Figure 1. The length of the coaxial feedline is unimportant, being long enough to reach comfortably from antenna to ham shack.

The Scotsman's Dipole

The basic dipole is about the least expensive antenna an amateur can construct. It requires no antenna tuner and is made up of inexpensive components. A high frequency dipole with dimensions for the popular ham bands is shown in the illustration. With the exception of 160 and 80 meters, the dipole may be cut to length for mid-band operation, and will work well over the entire band. On the two lowest ham bands, however, different dimensions are required for the low and high frequency ends of the bands, since the width of these bands is a large percentage of the mid-frequency design point (see Figure 1).

Constructing the Dipole

Antenna dimensions for each amateur band are shown in the tabulation. The flat-top sections may be No. 12 enamel copper wire, single strand. The 80 and 160 meter dipoles are quite long and should be made of "hard drawn" copper wire to reduce stretching and sagging to a minimum in heavy winds. Each end of the wires is cleaned thoroughly of insulation and the outer ends are looped through the insulator "eye", twisted back upon themselves and securely soldered. A *strain insulator* (egg insulator) is placed at the center of the antenna, with the wires passing through the overlapping holes, twisted back upon themselves, and soldered after the feedline is attached.

Preparing the Feedline

Chapter 5 stressed the importance of keeping water out of the interior of the coaxial line. If care is taken the line may be attached directly to the dipole antenna without danger of moisture entering the line. The use of RG-58/U line is suggested for transmitter power up to 500 watts, and the heavier RG-8/U line should be used for high power equipment. In the case of the RG-58/U, this is how you prepare the line:

The outer insulation is removed from the end of the coaxial line with a sharp knife, as follows. The insulation is carefully slit in a circle about 3 inches from the end of the line. Next, the jacket is slit

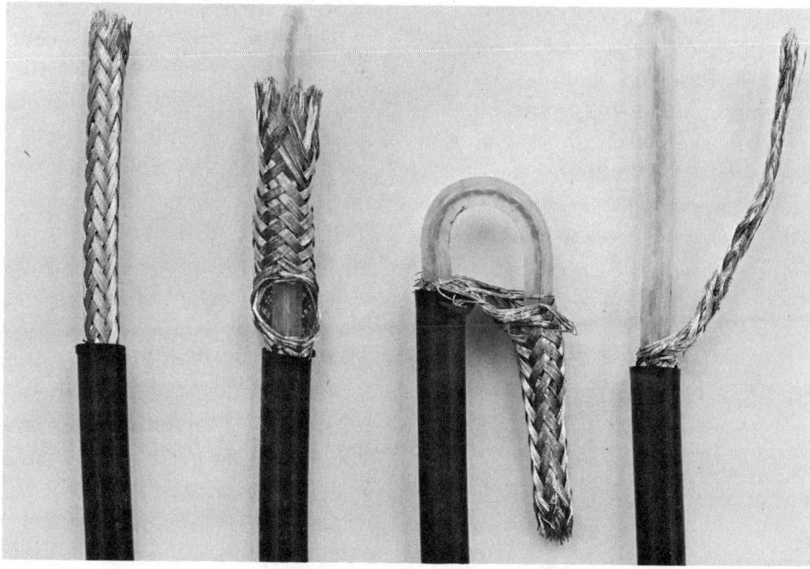

Fig. 2 A PIG-TAIL LEAD MADE THE EASY WAY. Left, the outer jacket is removed from the coaxial cable. Next, a hole is picked in the outer braid just above the jacket. The inner conductor is now carefully fished through the hole (center). Right, the completed pig-tail lead ready for use.

from this circular cut toward the end of the cable, as described in Chapter 5. The slug of vinyl insulation is removed, exposing the outer, braided conductor of the line. The braid is pushed back upon itself to loosen the weave. A sharp tool, such as a nail or awl, is used to open up a hole in the braid about a half-inch from the insulating jacket. The fine wires are separated with care, using the point of the tool, until a hole nearly equal in diameter to the diameter of the cable is made (Figure 2). Now, the cable is bent sharply at the hole and the insulated inner conductor is "fished" through the hole you have just made, using the nail or awl as a sort of hook. Once the inner conductor has been carefully pulled through the hole in the braid, the empty braid may be flattened into a pig-tail lead, ready for connection to one side of the dipole antenna. The last step is to strip the insulation from the end of the center conductor and *tin* the conductor (coat it with solder). The pig-tail lead is soldered to one half of the dipole and the center conductor is soldered to the other half. The weight of the line should be supported by the braided pig-tail rather than by the more fragile center conductor. The exposed end of the line is now firmly wrapped with black vinyl electrical tape and given a good coat of moisture-resistant sealant, such as *General Electric RTV-102.* This material is a rubbery gook, resembling bathtub calk (which also does a good job of sealing antenna connections). You can buy either sealant in a plastic tube at most hardware stores. Do a good job to make sure no pinholes exist in the seal to let water enter the line. Water soaked line makes for very weak signals! Finally, cut the line to length, pass it through the wall of the house, place a coaxial plug on the station end of the line and you are ready to erect and use the antenna. Excess cable may be coiled up in your operating room before the plug is attached firmly to your transmitter.

Erecting the Dipole Antenna

The dipole can be supported at its operating height by ropes holding the end insulators. Trees will come in handy at this point, as will a small mast placed atop your home. The dipole should be 30 to 50 feet high for best results--the higher the better! And as much in the clear as possible. Don't worry if the dipole is not "aimed" in the direction you wish. The figure-8 pattern is very broad and you can work nearly all around you, except off the very ends of the antenna. Height is more important than direction. If you do have a choice of direction, run the dipole in a north-south direction (in the U.S.) for maximum DX coverage.

The Multi-Band Dipole Antenna

The simple dipole is a great performer on a single amateur band. It is a knotty problem, however, to erect more than one dipole antenna when you live on a small lot, or plan to erect various antennas for more than one band. A *multi-band dipole* antenna might be the answer to your problem. This device is capable of working on two or more ham bands and requires only a single feedline. Several designs are discussed in this chapter and one of them may be just the antenna for you!

It must be remembered that because a multi-band antenna works on several bands it is capable of efficiently radiating transmitter harmonics on the higher bands. Use of a multi-band antenna should be tempered with caution since some transmitter designs emit strong harmonic signals. Use of a suitable harmonic filter or antenna tuner with *any* multi-band antenna is a good idea. More about this important point later in this Handbook.

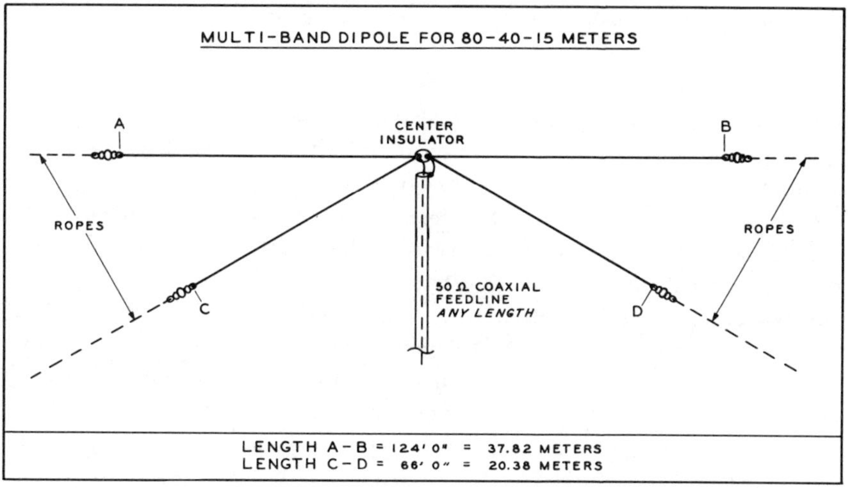

Fig. 3 **PARALLEL CONNECTED DIPOLES** allow two band operation. Ends of the dipoles should be separated three or four feet. Dipoles are connected at the center insulator and fed in parallel by a single transmission line. Forty meter dipole works on the third harmonic for 15 meter operation. Maximum dipole radiation is broadside to wires (into and out of the page).

A Multi-Band Dipole for the 80, 40 and 15 Meter Novice Bands

A multi-band dipole suitable for operation on 80, 40 and 15 meters is shown in Figure 3. Two separate dipoles are joined at the center insulator and fed in parallel by a single coaxial line. The 80 meter dipole functions on that band, and the 40 meter dipole serves double duty by working on 40 meters in normal fashion and also operating on 15 meters, on the third harmonic frequency of the antenna. Only that particular dipole functions which is excited by energy at or near its normal resonant frequency.

Constructing the Antenna

The four wire sections of the antenna flat-tops are cut to length, allowing enough extra wire at each end to loop around the insulators. Scrape the enamel insulation from the ends of the wires and fasten them securely around the insulator "eyes", soldering all joints after the antenna is completely assembled. It's a good idea to lay out the antenna and wire it up in a clear space to keep the builder from creating a rat's nest, with himself in the middle!

Erecting the Antenna

Once the double dipole assembly is completed, the coaxial line is attached to the center insulator, bearing in mind the remarks made previously about water-proofing this all-important joint. The complete antenna is supported by the 80 meter wires, with the 40/15 meter dipole draped beneath the longer wires. Spacing between dipoles is not critical and may be two to five feet at the ends. In fact, the dipoles can run at right angles to each other if that arrangement fits the available space more conveniently. Each individual dipole, however, should run in a straight line, if possible. Place the antenna as high in the air as you can. Operation on any band is automatic--no changes or adjustments need be made to the antenna when changing bands.

In case loading difficulty is experienced on one band, adding a few feet of coaxial line at the station end usually will overcome the difficulty. A short length of line (with fittings) and a coaxial splice adapter (*Amphenol* 83-1J for RG-8/U or *Amphenol* 31-219 for RG-59/U) will do the job in no time at all.

* * *

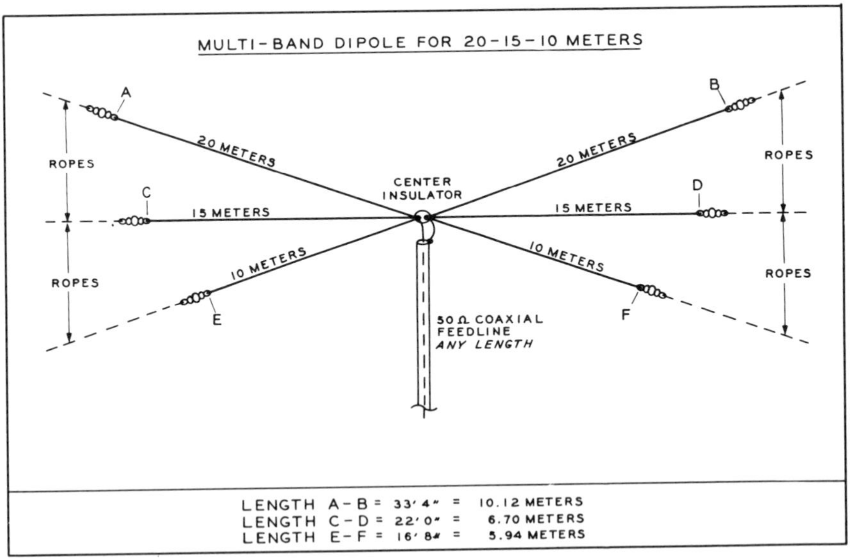

Fig. 4 TRIPLE DIPOLE FOR 20-15-10 METER OPERATION. To reduce interaction the 15 meter dipole should be run at right angles to the other two antennas.

A Multi-Band DX Dipole for 20, 15 and 10 Meters

The three popular DX bands may be covered by a 3-band dipole that provides excellent operation. Erected high in the clear, this DX antenna will be worth its weight in exotic QSL cards!

Again, the parallel principle of dipole operation is used, three separate antennas fed by one transmission line. The multi-band assembly is built in the same manner as the dipole previously described for the 80, 40 and 15 meter bands, using the dimensions shown in Figure 4. Assembly of the antenna is quite similar to that described for the previous antenna.

It is suggested that the 15 meter dipole antenna be placed at right angles to the others to reduce electrical interaction between the antennas to a minimum. For convenience, the 15 and 10 meter dipoles may be dropped down at angles from the 20 meter section, as is done in the so-called "Inverted V" antenna, to be described later in this chapter.

Trap Dipole Antennas

Parallel connected dipole antennas are simple antennas to get working, but many amateurs prefer a *trap dipole* for multi-band operation as only one flat-top wire is required. A trap dipole operates on the principle of parallel tuned switching circuits placed at critical points in the antenna element. The circuits, or traps, electrically connect or disconnect the outer sections of the element as the antenna's excitation frequency is changed.

At the lowest operating frequency, the traps have a minimum effect upon the antenna, which is resonant at a frequency determined by its overall electrical length, including the effect of the traps. At some higher frequency, the traps are parallel resonant, effectively disconnecting the end sections of the antenna. The length of the center section only is therefore resonant at the highest operating frequency, and the complete antenna assembly is resonant at the lowest operating frequency (Figure 5).

On the lowest band, the traps contribute to antenna length and make the overall antenna length somewhat shorter than an equivalent dipole for the lowest band. The trap principle may be used to make five, four, three, or two-band dipoles. A few of the more popular and proven multi-band versions are discussed in this chapter.

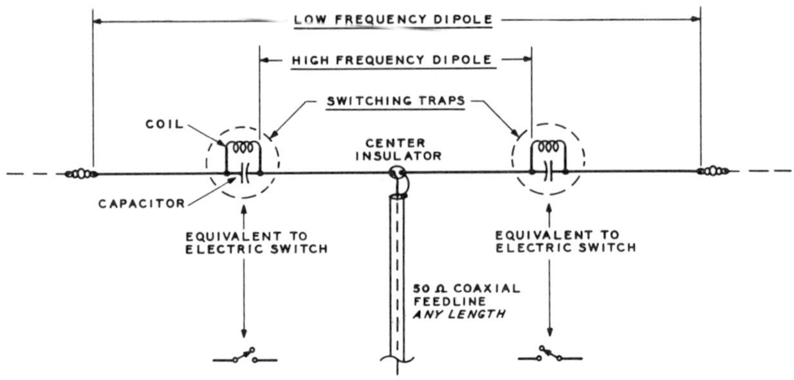

Fig. 5 MULTIBAND TRAP DIPOLE provides operation on different ham bands with single antenna wire. The center section of the antenna is a simple dipole at the highest band of operation. The parallel tuned trap circuit is resonant near this frequency and acts as an open circuit. At the lowest band of operation, the trap circuit acts as an inductance, effectively completing the connection between the end sections of the antenna and the middle section. Because of the loading effect of the traps, overall length is shorter than normal for the lower frequency band.

A Two-Band Trap Dipole for 20 and 15 Meters

The dimensions and trap data for a simple two-band trap dipole for 20 and 15 meters are shown in Figure 6. Trap adjustment is discussed later in this chapter. The 15 meter center section of this two-band antenna has a normal length, but the overall 20 meter length is shortened by the loading effect of the traps. The traps have an "electrical length", plus a physical length of about two inches, and the sum of these lengths must be subtracted from the overall 20 meter length.

Construction of a typical trap is shown in Figure 7. The trap is built around a strain insulator which removes the antenna tension from the coil and capacitor. The insulator itself has a small amount of capacitance so it must be considered as a part of the trap circuit. The coil may be placed around the insulator or beside it, with the capacitor parallel connected across the coil. To attach the trap to the wires of the antenna, two inch leads are soldered to the end connections of the trap. The leads may then be tinned, twisted around the antenna wires and soldered, as shown in the photo.

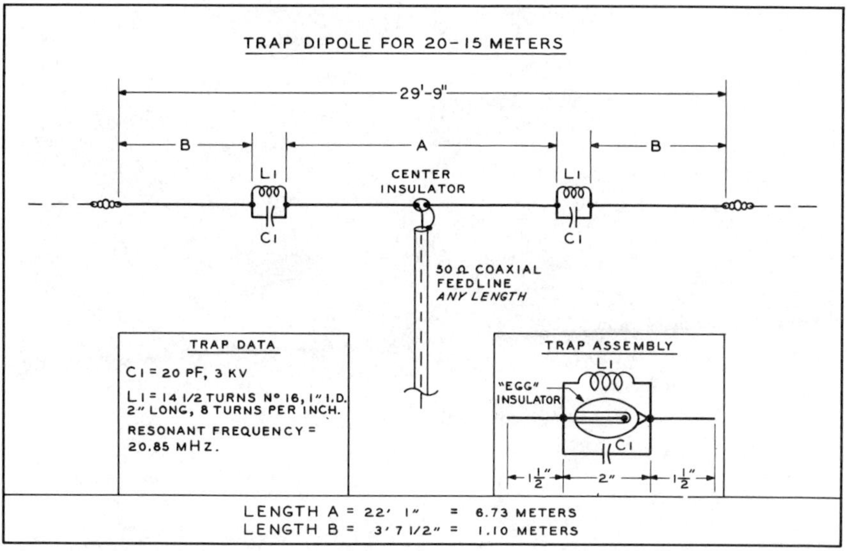

Fig. 6 TWO BAND DIPOLE has traps tuned to high frequency band. Trap assembly is shown in Fig. 7. Inductors are made of prefabricated coil stock. Fifteen meter resonance is determined by length of inner section and 20 meter resonance is determined by length of end sections. Traps are pre-set to proper frequency before installation in antenna.

Assembling and Tuning the Traps.

Attach a foot of wire to each end of the strain insulator, looping the wire through the insulator and fastening it back upon itself. Cut the wire and solder the joint. The coil is then slid over the insulator, or placed beside it. Solder the coil leads to those of the insulator, then connect the capacitor in parallel with the coil and insulator, as shown in the photograph.

After the trap is assembled, place it in a clear spot atop your desk or workbench, away from metal objects. Loosely couple it to a grid dip oscillator. Check the trap resonant frequency against a calibrated receiver. Make the measurement several times so as to pin-point the frequency of the trap as accurately as you can. The exact resonant frequency of the trap isn't too important, as long as it occurs near the low frequency end of the band, as specified in the illustration. Both traps in a set should have a resonant frequency within about 50 KHz of each other. Frequency adjustment may be achieved by removing (or adding) turns to the coil, a fraction of a turn at a time.

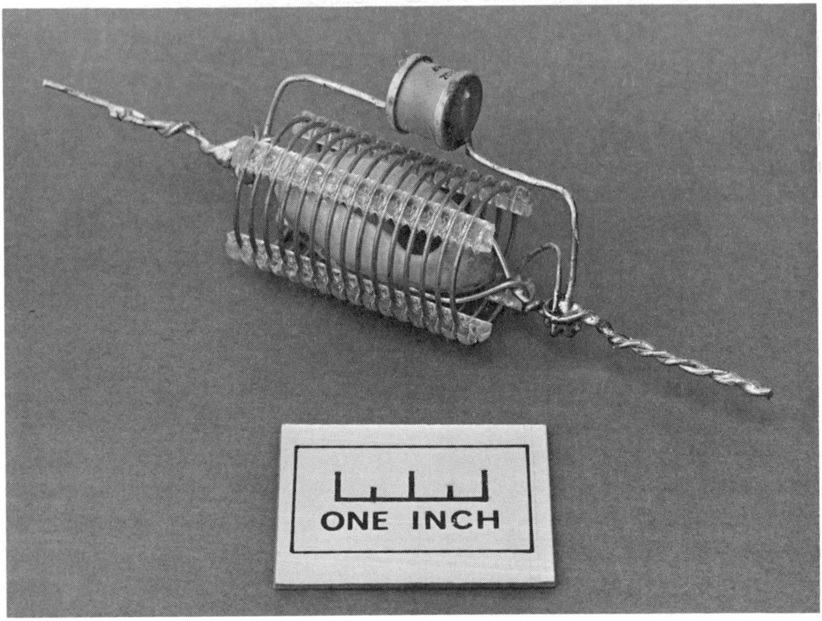

Fig. 6 TYPICAL TRAP ASSEMBLY. Trap is built around a ceramic strain insulator. Tinned wire leads are attached to the insulator and an airwound coil is slipped over it. Coil leads are soldered to insulator wires. Centralab type 850 or 855 ceramic transmitting capacitor is connected across coil. Resonance is established by trimming coil a quarter-turn at a time, checking resonant frequency with grid-dip oscillator.

Antenna Assembly

Once the trap is tuned to the correct frequency, it may be soldered into the antenna assembly. It is suggested that the antenna be assembled at waist height, stretched between two accessible points for ease of construction. The last step is to attach the transmission line to the antenna and the electrical work is done.

The last important consideration is moisture-proofing the traps, which is really more of a problem than building them! The traps may be used without a protective shield if it is guaranteed not to rain. A temporary rain shield may be made by enclosing the traps in kitchen plastic sheet, or "Baggies", and taping the plastic closed around the trap. A far better and more permanent weather shield may be made from a polyethylene "squeeze bottle", obtainable at a drug store. The best

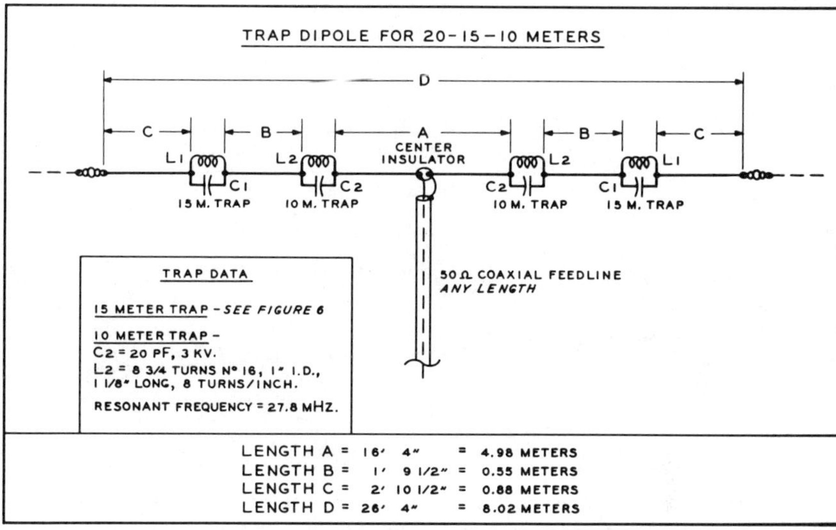

Fig. 8 POPULAR TRI-BAND DIPOLE permits operation on three DX bands. This design is used in most tri-band Yagi beam antennas. Traps are pre-set before installation in antenna and require no further tuning.

protective shield (and most difficult to construct) is made of plastic tubing with end pieces cemented on to form a waterproof cylinder surrounding the trap. Regardless of the waterproofing scheme, it is imperative that the traps be protected from water, otherwise, the circuits will become detuned when wet and may possibly flash over and destroy themselves during operation.

A Tri-Band Trap Dipole for 20, 15 and 10 Meters

The trap technique may be extended to three bands, as shown in Figure 8. Two sets of traps are required, one for 10 meters and one for 15 meters. While the 10 meter dimensions of the inner section of the dipole are normal, the overall antenna lengths for the 15 and 20 meter resonant sections are shortened by insertion of the traps. Overall antenna length is only 26'4''. Construction and adjustment of the traps is as discussed in the previous section of this chapter.

A Five-Band Trap Dipole for 80, 40, 20, 15 and 10 Meters

The ultimate in trap dipole design is shown in Figure 9 for a five-band antenna assembly. Because of the unique design, only one set of

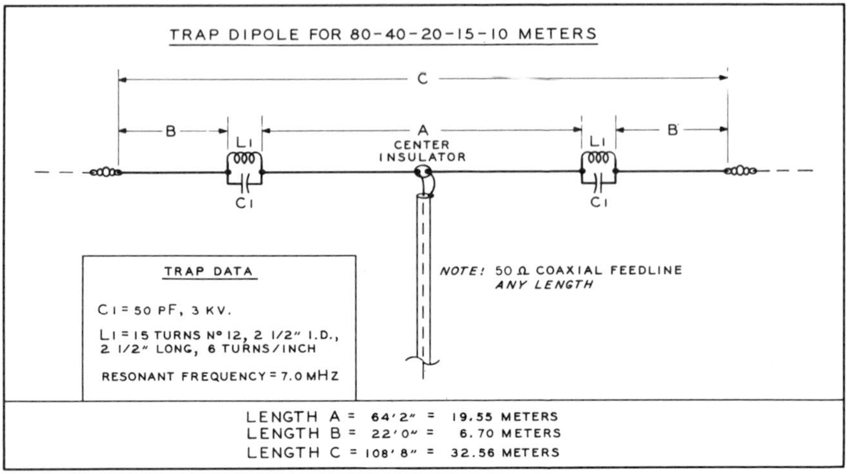

Fig. 9 NOVEL MULTIBAND DIPOLE PROVIDES operation on all popular h-f bands. Center section functions as a dipole on 40 and 15 meters and complete antenna is resonant in the 80 meter band. Traps also provide proper reactance so as to resonate center section for proper operation on 20 and 10 meters.

traps is used, resonant near the 40 meter band. On 80 meters, the whole antenna system is resonant at a center frequency of about 3.8 MHz. The center section is resonant at 40 meters and operates on the third harmonic for 15 meter service. On 20 and 10 meters, the 40 meter traps provide sufficient reactance so as to resonate the center section of the antenna for proper operation on each band.

This antenna, as well as the other multiband antennas shown in this chapter, do not require the use of a balun, however one may be used if desired. Adding a balun tends to isolate the outer conductor of the coaxial line from antenna currents, reducing line radiation. In most cases line radiation is no great problem, but it can cause TVI and possibly feedback in your transmitting equipment, particularly on 10 meters. In any event, adding the balun will not detract from proper operation of the antenna.

A Four-Band Trap Dipole for 40, 20, 15 and 10 meters

A five-band dipole assembly may be simplified and repackaged to fit a smaller space if 80 meter operation is omitted from the design. In addition, the four-band antenna may be fed with a normal 50 ohm coaxial transmission line. The four-band trap dipole is shown in Figure 10. The antenna is resonant near the mid-points of each amateur band yet requires only two traps for proper operation.

Trap Antenna Adjustment

Proper operation of any trap dipole antenna may be determined with the aid of an SWR meter. Typically, the antenna should exhibit an SWR of about 1.5 or so at the resonant frequency, rising to around 2 or 2.5 at the edges of each band. (Bandwidth of the 80 meter, five-band trap dipole is restricted to about 100 KHz on the 80 meter band, but is normal on the higher bands).

While not absolutely necessary for good operation, it is recommended that a balun be placed between the trap antenna and the feedline to permit more accurate SWR measurements to be make. If it is desired to shift the center point of any segment of the antenna, it is done by changing length of the antenna sections, and *not* by readjusting the traps. Adjustment, if any, should be done to the highest frequency antenna first, followed by the next lower frequency segments. It must be remembered that adjustments made to one antenna section affect the adjustments of the lower frequency segments, since the higher frequency segments form a part of the lower frequency ones.

Don't deliberately look for trouble, because the measurements given for these antennas have been tried and proven in many installations with complete success -- and loud signal reports!

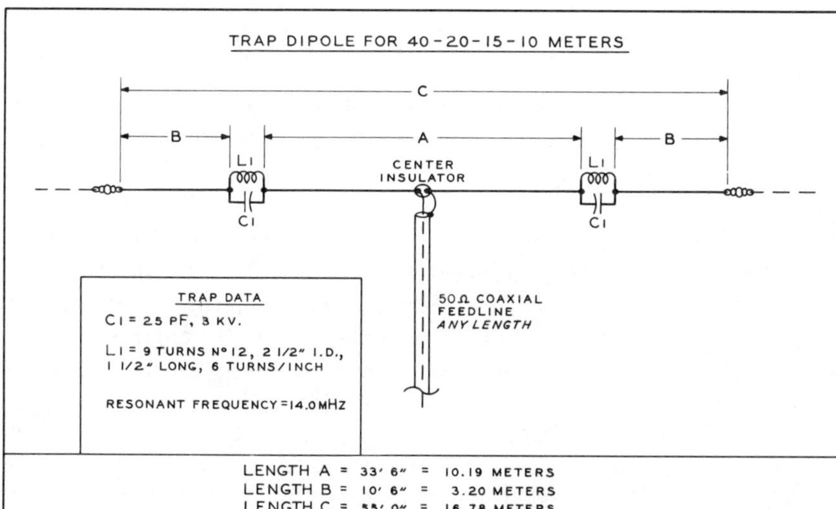

Fig. 10 COMPACT MULTIBAND DIPOLE ANTENNA FOR 40, 20, 15 and 10 METERS. Only about 55 feet overall, this simple multiband antenna functions on four DX bands. A balun may be used at the center insulator, if desired. The antenna should be strung in a straight line for best results.

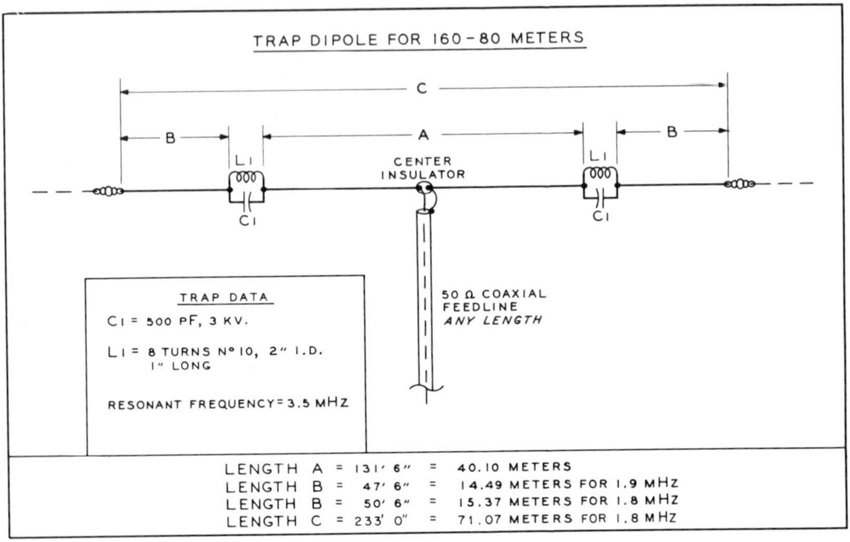

Fig. 11 UNIQUE TRAP DIPOLE FOR 160 and 80 METERS provides strong signals for trans-Pacific DX tests. Designed for operation in the lower portion of the 160 meter band, the tip sections may be shortened for operation in the 1.9 MHz segment of the band. Overall length will then be about 227'6''.

A Two-Band Trap Dipole for 160 and 80 Meters

The low frequency enthusiast who works both 80 and 160 meters will appreciate this two-band antenna designed by JH1LKH of Japan who has used it with success on trans-Pacific 160 meter DX tests. The assembly is shown in Figure 11, with dimensions given for both the low and high frequency segments of the 160 meter band. The 80 meter section is cut to a center frequency of 3.5 MHz for CW DX operation.

The 80 meter resonant frequency may be changed by varying the length of the center portion of the antenna, while the length of the tip sections may be altered to move the center frequency about in the 160 meter band.

Since the antenna height will probably be small in terms of wavelength (very few hams can get their 160 meter antenna a half-wavelength in the air!), the radiation resistance at the center feed point will probably be quite low. This means the bandwidth of the antenna on 160 meters will probably be only about 100 KHz and the SWR will run close to 1.7 or so at the center frequency. This works no hardship, as most modern equipment will accept this value of SWR with ease.

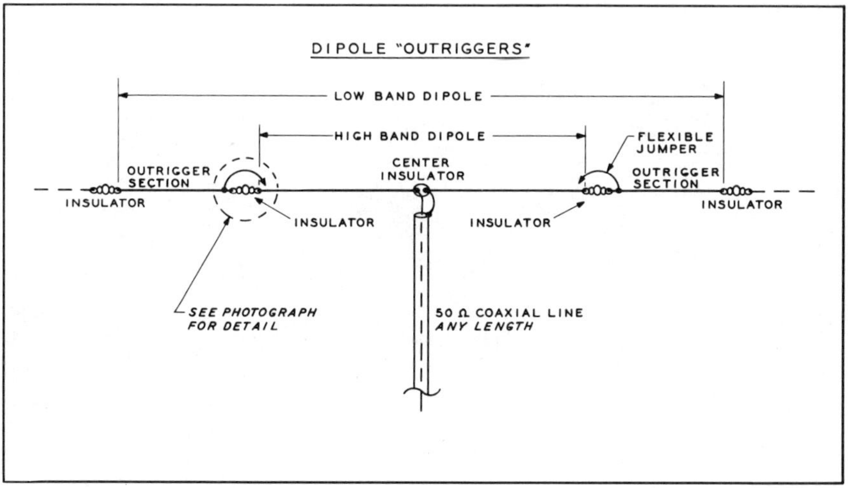

Fig. 12 DIPOLE ANTENNA MAY BE OPERATED ON LOWER FREQUENCY BAND by addition of "outrigger" sections to ends of flat-top. Overall length of dipole plus outriggers is resonant near center of low band allowing two band operation.

The trap capacitor is made up of five 100 pF ceramic units connected in parallel across the coil. Coil turns are adjusted until the trap resonates at about 3.5 MHz.

Operating Your Dipole on a Lower Frequency Band

A simple dipole antenna may be operated on a lower frequency band by adding "outrigger" sections to the ends of the flat-top. The outrigger wires merely add length to the dipole and re-resonate it to a new, lower frequency. The outriggers are attached to the antenna tips by means of copper battery clips as shown in Figure 12. When unclipped, the outriggers merely serve as support wires for the dipole.

The outriggers are short sections of copper wire with insulators at the far ends. To use the antenna at the lower frequency, the dipole is lowered and the outriggers clipped across the end insulators. To retune to the higher frequency, the outriggers are disconnected. In this way, the 20 meter antenna may be made to work on 40 or 80 meters, or *any* dipole may be made to work at a lower frequency than that for which it was designed. A dipole cut for 80 meter phone, for example, may be made to work at the low end of the 80 meter CW band, by the addition of suitable outriggers.

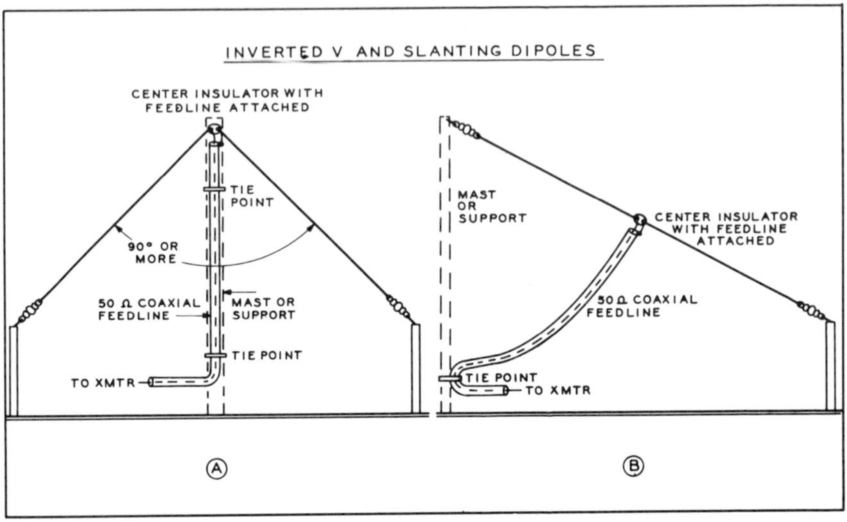

Fig. 13 **DIPOLE MAY BE INVERTED OR SLANTED** as shown with no ill results. Inverted V dipole is very popular on the low bands as only one support is needed and high current point of antenna is elevated.

If the outer ends of the dipole are tilted towards the ground, it may be possible to reach the outrigger clips from a small step ladder, thus eliminating the necessity of raising or lowering the antenna each time it is desired to jump to the lower frequency band.

Of course, series-connected multiple outriggers may be used, if desired, for operation of an antenna on two or more lower frequency bands.

The Inverted V and Slanting Dipole Antennas

The popular *Inverted V* antenna is merely a dipole supported at the center with the flat-top wires drooping towards the ground. It may be supported from a single centerpole, with the dipole wires used as guy wires for the pole. Thus, it is a good antenna to use in a restricted space, as end supports are not needed. Its operation is equivalent to that of a horizontal dipole; any dipole antenna described in this section may be transformed into an Inverted V by merely drooping the tips down, as shown in Figure 13. The included angle of the V should be as large as possible, and not less than 90 degrees, as shown in the illustration. Inverted V's are good DX antennas particularly on the lower frequency bands where beams are few and far between.

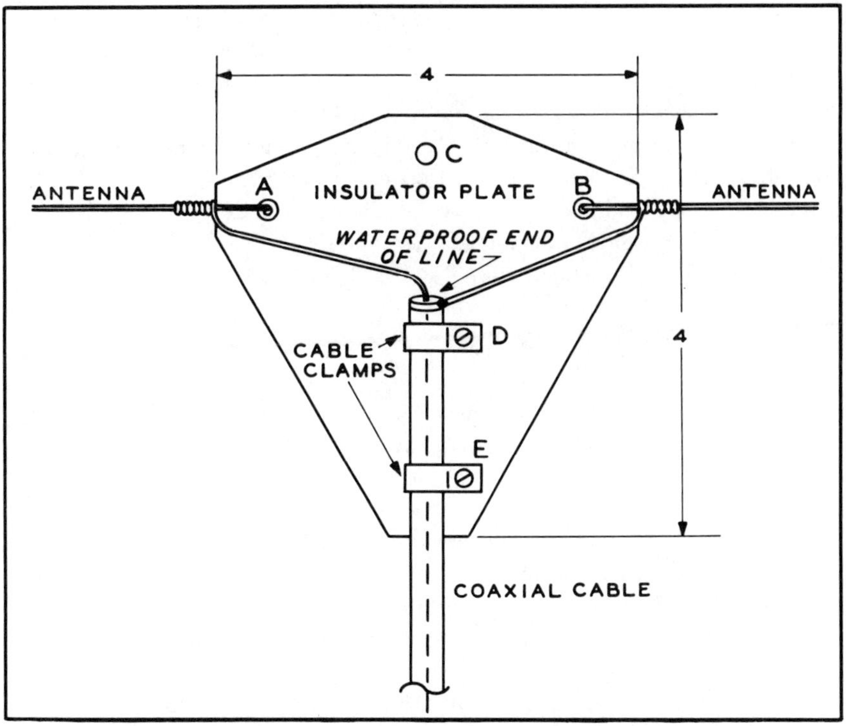

Fig. 14 CENTER INSULATOR FOR DIPOLE OR INVERTED V ANTENNA is made of insulating material. Cable clamps remove tension from antenna connections.

A Rugged Center Insulator You Can Build

Make the center insulator shown in Figure 14 and your job of constructing a doublet antenna will be much easier! This useful and inexpensive insulator provides convenient tie points for the antenna wires and also for the coaxial transmission line.

The insulator plate is cut from a 1/4-inch thick piece of insulating material such as *micarta,* lucite or plywood. If wood is used, it should be given two coats of waterproofing shellac. Size of the plate is not critical.

Holes A and B are drilled to freely pass the antenna wires and hole C is used to support the dipole at the center, as is done with an inverted V antenna.

Two cable clamps hold the coaxial feedline firmly to the insulator. Sturdy metal clamps, such as those used in automobiles to fasten the hydraulic lines in place are recommended. The clamps are held in position by bolts passing through mounting holes D and E. These holes are drilled off-center so that the coaxial cable drops down the center line of the insulator.

After the insulator is cut out, the edges of holes A and B should be countersunk a bit with a large drill so as to provide a rounded edge for the antenna wire, otherwise it may be chafed at these points. The last step is to thoroughly waterproof the end of the coaxial line, as explained earlier in this Handbook.

The Twin-T Antenna for 40 and 80 Meters

The Twin-T antenna is a center-fed dipole system designed for 40 and 80 meter operation. It requires no switching or tuning and is fed with a 50 ohm transmission line. Antenna polarization is vertical on 80 meters and horizontal on 40 meters. The antenna is compact enough to fit on a small lot and requires no trap circuits to adjust. Best of all, the Twin-T can be built in a day and the cost is relatively modest.

A complete Twin-T antenna system is shown in Figure 15. It is composed of a two wire flat-top fed at the center of one wire with a vertical length of open wire "TV-type" two conductor transmission line. The vertical section is fed at the bottom with a random length of

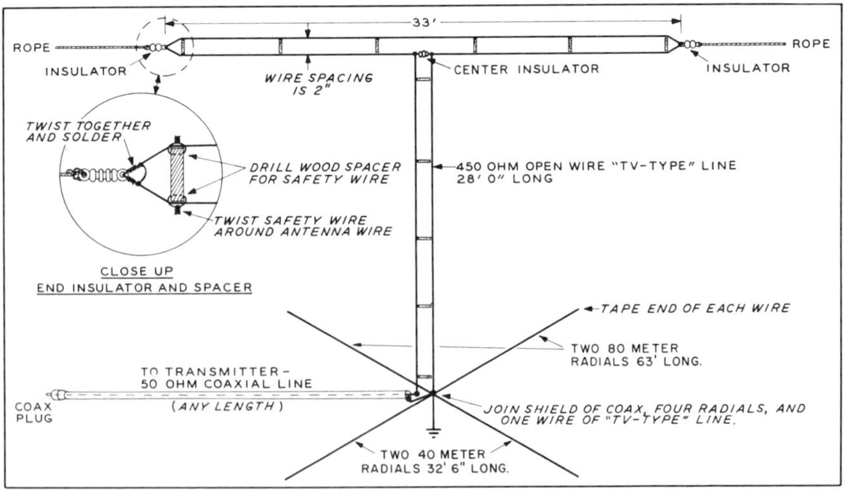

Fig. 15 TWIN-T ANTENNA FOR 40 AND 80 METERS. Vertical section may be made of two #16 enamel wires spaced 1½'' apart using wood spacers placed two feet apart, such as described for the horizontal dipole section.

50 ohm coaxial transmission line. Two or four radial ground wires are placed at the bottom of the vertical section, as shown in the illustration.

On 40 meters, the horizontal portion of the Twin-T antenna acts as a dipole with the ends bent back upon themselves, reducing the overall length of the dipole to one quarter wavelength. The ends of the dipole present a very high impedance to the open wire transmission line which acts as a matching transformer to step the dipole end impedance down to match the characteristic impedance of the transmission line.

On 80 meters the Twin-T antenna resembles a top loaded vertical antenna, fed in one leg. The loading is provided by the horizontal portion of the antenna which does very little radiating. The step-up in impedance provided by the vertical wires provides a close match between the antenna impedance and the transmission line. Resonant radial wires for each band are placed at the base of the vertical section of the antenna.

Antenna Assembly

Two #12 enamel wires 34'6'' long are cut for the flat-top section. Cut one wire at the center, scrape all ends carefully with a knife blade to remove the enamel and attach insulators at the ends of the antenna and also at the center break. Overall antenna length should be exactly 33 feet. Make sure the middle insulator falls at the exact center point of the wires. If you do this job with the antenna temporarily slung between two handy supports it will be a simple and uncomplicated task.

The next step is to make up six spreaders that will hold the flat-top wires at the desired 2 inch separation. The spreaders can be made of lucite rod or wood blocks. The spreaders measure about 3'' x 1'' x ½''. Wood spreaders should be given a coat of waterproofing shellac or varnish. Drill the ends of the spreaders to pass a safety wire.

The blocks are placed in position between the antenna wires and are lashed in place with the short safety wires. Stretch the antenna tight between the supports and solder all the antenna joints.

The last step is to cut the open wire TV line to the proper length and solder the end wires across the center antenna insulator. The antenna is erected in the air and the coaxial line and radial ground wires are connected at this time.

Forty meter resonance can be changed by varying the length of the flat-top and 80 meter resonance changed by varying the length of the vertical wires.

Dipole Antenna Placement

Where to place the dipole antenna once it is completed and ready for use? There's no specific answer, but here are some suggestions which will aid you in getting the best results from your antenna installation.

Get the antenna in the air as high as possible for best results. Antenna heights up to 150 feet or so are used by some ardent DXers! The average amateur, however, usually has to settle for something less than this stratospheric elevation. Generally speaking, a minimum dipole antenna height of about 30 feet above the ground is acceptable for all-around operation on any high frequency band, providing the antenna is reasonably in the clear. For better results for long distance DX work a height of 45 feet or so is recommended. Those lucky amateurs able to erect their antenna 60 or 70 feet in the air will get even better results, although the additional height is not as striking as far as signal reports go as one would suppose.

Antenna height achieves one very important objective: it removes the antenna from the immediate vicinity of power lines, house wiring and nearby metallic structures. Separation from these objects tends to improve the antenna radiation pattern for DX and to reduce noise pick-up on reception. Regardless of the antenna height it is not a good idea to run the antenna parallel to power or telephone lines. Proximity to such wires can lead to unwanted television interference, telephone interference and noise pickup on reception. Try and keep your antenna away from elevated wires and place it at an angle to them to reduce coupling.

In summary, the dipole antenna should be as high and as in the clear as possible, should not run parallel to utility lines and -- for all-around DX operation -- should run in a generally north-south direction so the main radiation lobes of the antenna cover the areas of the world having the greatest radio amateur population.

Warning! No antenna should cross over a utility or telephone line because of the danger of shock or electrocution if the antenna wire touches the "hot" utility line. Amateurs have been killed trying to pass the antenna wire over a power line, and deaths have been reported of hams electrocuted when their antenna dropped on a power line during a wind storm. Play safe! Keep your antenna *away* from utility lines of all types!

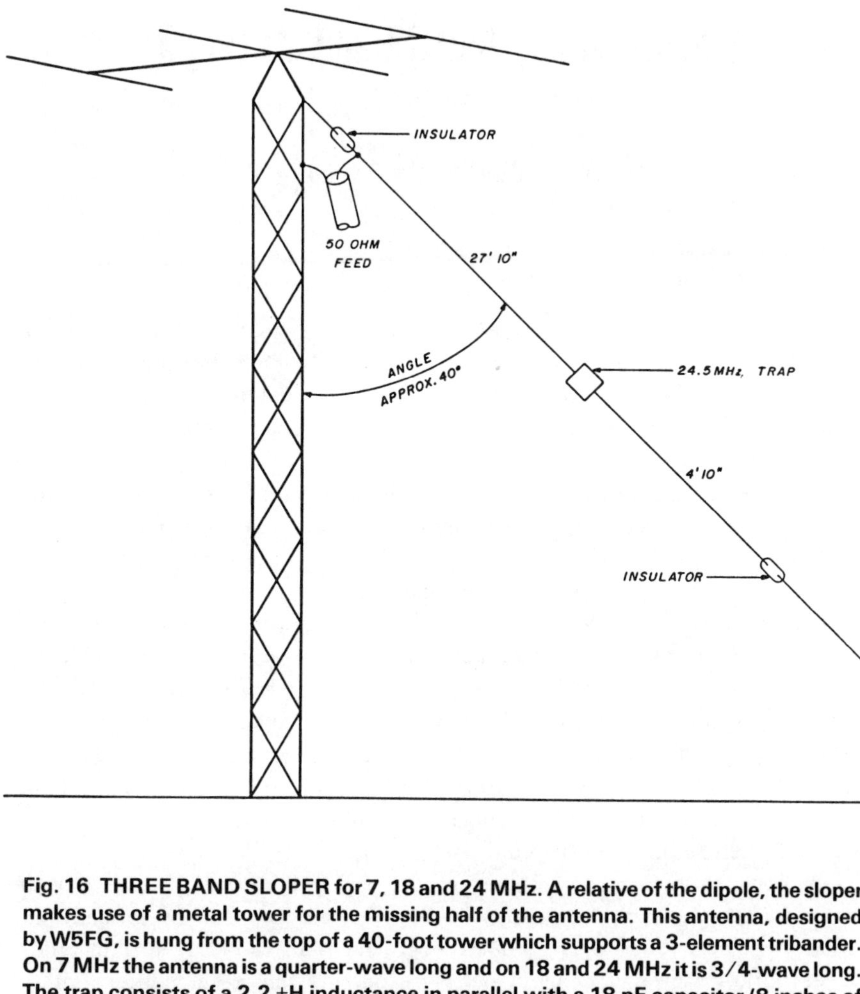

Fig. 16 THREE BAND SLOPER for 7, 18 and 24 MHz. A relative of the dipole, the sloper makes use of a metal tower for the missing half of the antenna. This antenna, designed by W5FG, is hung from the top of a 40-foot tower which supports a 3-element tribander. On 7 MHz the antenna is a quarter-wave long and on 18 and 24 MHz it is 3/4-wave long. The trap consists of a 2.2 +H inductance in parallel with a 18 pF capacitor (8 inches of RG-58/U). The trap is adjusted to 24.5 MHz with a dip oscillator before it is placed in the antenna. The inductor is 10 turns of #14 wire wound on a 1-5/8 inch diameter form. Adjusting the slope angle and pruning the outer section will bring the SWR to 1.5 or less across all three bands.

Chapter 8

The End-Fed, Multi-Band Antenna

(Or, When in Doubt, Use a Long Wire)

The simplest, least expensive and often the most practical amateur antenna for multi-band operation is a random length *Marconi antenna*. The famous experimenter Guglielmo Marconi used an antenna of this type in his early work, hence the name. This uncomplicated "sky wire" is also called an *end-fed* antenna, a *directly-fed* antenna and sometimes a *long wire* antenna. Don't loose any sleep over the names as they all refer to the same thing--the single wire Marconi antenna shown in various forms in Figure 1.

The smallest and most compact end-fed antenna is the popular coil-loaded whip that mobileers use for 80 meter operation. A large portion of this mini-antenna is wound up in the loading coil required to tune the antenna to resonance. The flexible 10 meter whip is also an end-fed antenna, as is the 1000 foot long wire antenna used by some 20 meter DXers. Regardless of antenna length any antenna that is fed at one end instead of at the middle is classified as an end-fed antenna. This chapter describes several interesting end-fed antennas and one of them may just be the antenna for you!

A critical and important part of the end-fed antenna is the *ground* connection. The mobileer uses the body of his car for the ground and some over-optimistic amateurs drive ground rods into the soil in order to establish a ground connection. More on this important subject later.

Matching the End-Fed Antenna

The tuning of the end-fed antenna, as well as any other antenna, depends upon matching the ratio of the voltage to the current at the feed point to the transmitter, as explained in Chapter 3 of this Handbook. This ratio of voltage to current is called the *radiation resistance*

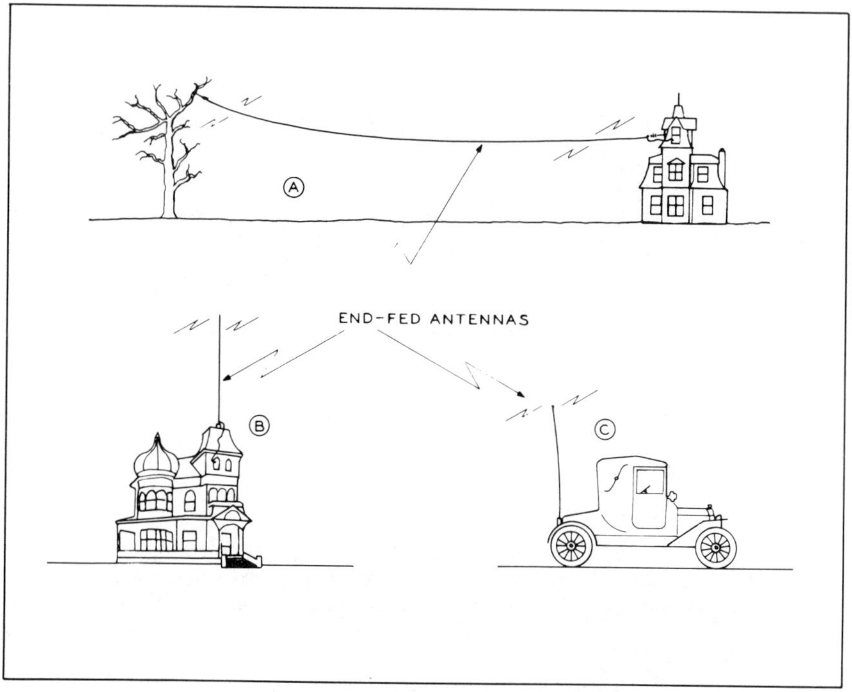

Fig. 1 END-FED ANTENNA TAKES MANY FORMS. Marconi antenna, used with a ground or radial ground wire is called an end-fed antenna. A- End-fed long wire is popular h-f antenna. B- Vertical or ground plane is a form of end-fed antenna. C- Short mobile whip is also end-fed antenna. These are all versions of the same antenna configuration!

of the antenna. In the case of a random length, end-fed antenna, the radiation resistance at the feed point can vary over a large range, depending upon the operating frequency and the actual length of the antenna. The measured radiation resistance can be as high as 5000 ohms or as low as 2 or 3 ohms, and the antenna can exhibit reactance which (as explained in Chapter 3) makes it difficult to load.

To compensate for the unusual values of radiation resistance and reactance presented to the transmitter by the end-fed antenna, an impedance matching device called an *antenna tuner* must be used. The tuner matches the antenna to the 50 ohm coaxial output termination used by most of today's radio equipment. Use of an antenna tuner, plus an SWR meter permits the operator to quickly transform the radiation resistance of his end-fed wire to a value which will efficiently match his transmitter. He does this by adjusting the tuner for minimum value of SWR as read on the meter of the instrument (Figure 2).

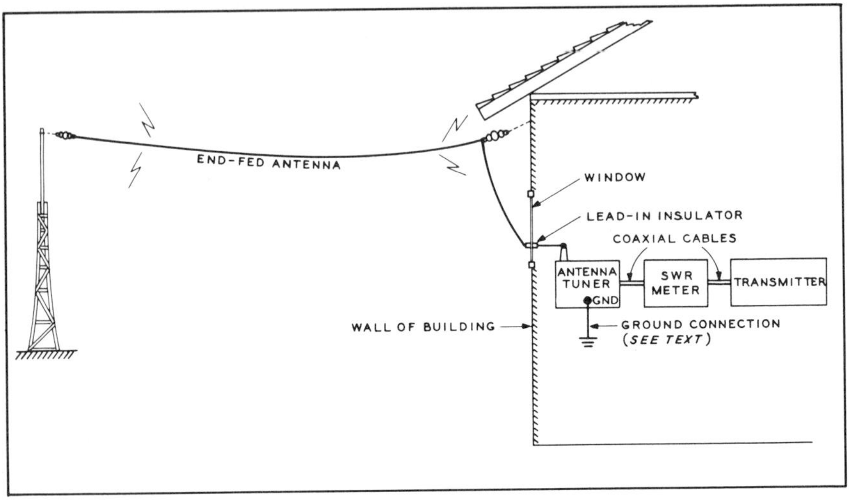

Fig. 2 RANDOM LENGTH END-FED ANTENNA and good ground connection, combined with antenna tuner and SWR meter, makes simple and effective antenna for all high frequency bands. The secret of success is efficient ground connection.

The All-Important Ground Connection

Many amateurs run into difficulty with the end-fed antenna because they have a poor ground connection, which can destroy the efficiency of the antenna and cause improper transmitter loading.

Experience has shown that a rod driven into the soil or a connection to a radiator or to a heating vent is a *poor* radio ground. To make matters worse, the longer the wire connection between the transmitter and the ground, the poorer is the electrical efficiency of the ground! To put it bluntly, many typical ground connections are worthless, especially on the higher frequency ham bands where the length of the ground lead is an appreciable fraction of a radio wavelength.

It is very difficult to make a good, low resistance ground connection. If the ground connection exhibits significant resistance to the flow of rf current, a portion of your power output will be lost in this resistance. In addition, if the chassis of your equipment is not at a ground potential, unwanted rf currents may flow on it, resulting in instability and audio feedback. In some instances, a poor ground connection results in a "hot" microphone that "bites" the operator whenever he touches it. In severe cases, actual damage may be done to the transmitter if an imperfect ground connection is used with an end-fed antenna system.

Broadcast stations have expensive ground systems that cover an area the size of a city block and some older shortwave point-to-point stations also have extensive ground systems. These elaborate installations are beyond the financial ability of most amateurs, and they must use a simpler and less expensive ground system to make their antennas work. Fortunately, this can be done by the use of *radial ground wires,* a modification of a more complicated grounding technique used in modern commercial shortwave transmitting installations.

In summary, then, the end-fed antenna has three important requirements that must be met to make it work effectively. These are:

1- A good electrical ground system is required.

2- An antenna tuner to match the end-fed antenna to the transmitter.

3- An SWR meter to determine proper adjustment of the antenna tuner.

The Radial Ground Wire

A few ready made radio ground systems exist and, if available, one of them should be used *in conjunction with the accessory radial ground wire.* An underground yard sprinkling system composed of copper pipe with soldered copper fittings may be used, provided the connecting lead from the pipes to the radio equipment is short and direct -- not more than a few feet long. A second available radio ground is the cold water distribution system of the home, provided it is composed of copper pipe having soldered joints. Water systems of iron pipe have questionable joints (as far as electrical conductivity goes) and plastic pipe water systems are useless as a radio ground.

Used in conjunction with a radial ground wire, either the sprinkling system or the cold water distribution system provides a good radio ground system for any radio antenna. The water system, or the radial ground wire by itself, is not as effective as the two used together. In any case, the accessory radial ground wire is the more important of the two ground connections and its use is mandatory for proper operation of the end-fed antenna system.

The radial ground wire (sometimes called a *counterpoise)* is an artificial electric ground that is very effective. It is simply an insulated wire, one quarter wavelength long at the operating frequency, connected to the transmitter at one end and run away from the equipment in a random direction, either indoors or outdoors. The far end of the wire

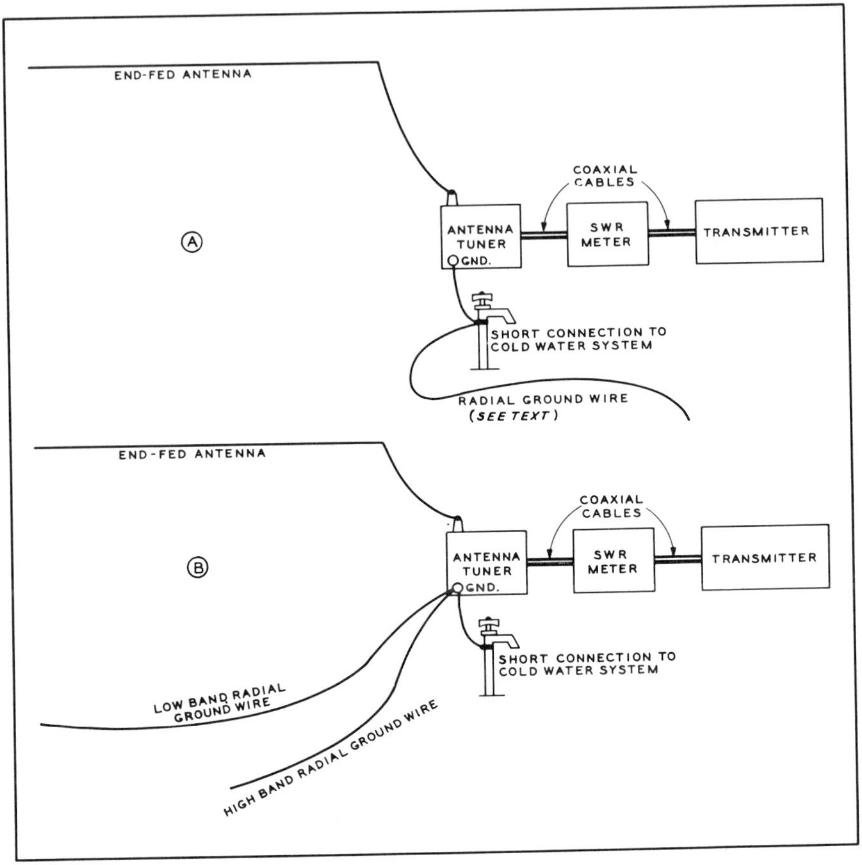

Fig. 3 MULTI-BAND OPERATION of end-fed antenna is possible when multiple radial ground wires are used along with good ground connection to cold water system of residence. Only a single wire is needed for one band operation (A) with multiple radials used for multi-band operation (B).

is left free. The station end is connected to the transmitter (Figure 3A). The end of the wire should be taped to prevent contact as it is "hot" with rf energy and can cause a nasty rf burn to anyone unfortunate enough to touch it when the transmitter is on the air.

As a radial ground wire is resonant, it will only work properly on one amateur band. Two or more radial wires may be attached to the transmitter for multi-band operation (Figure 3B). The radial placement is not critical, although it is usually run in a horizontal plane, along the floor of the radio room tacked against the wall, or perhaps out the window and along the side of the house. For the lower bands where the radial is quite long, it can be run through bushes or trees, a few feet above the ground. The radial wire should not actually touch the ground, nor any metallic object. It is not considered to be an antenna, so it does not have to be "in the clear", or positioned under the antenna itself, but it should be stretched out in as straight a line as possible, away from the transmitter. For portable operation, the radial wire can consist of a length of hookup wire, tossed along the floor behind the operating table and held in position with a book or two atop it.

If a reasonable ground connection is available, it should be used along with the radial. The combination of the best possible radio ground, plus a radial ground wire, "tames" the end-fed antenna and makes it a really effective antenna design for all-around amateur operation. Best of all, the end-fed antenna is unobtrusive and can be used in locations where a more highly visible antenna might bring trouble down upon the head of the amateur. A chart of radial wire lengths for the hf amateur bands is shown in Figure 4. For multi-band operation, a multiple radial may be made out of multi-conductor TV control cable, each wire of the cable being cut to the length of one radial for each band.

Bury Your Radial Wires?

Placing tuned radial wires about your property may be a real problem unless the wires are hidden in a flower bed or slipped among trees and bushes. In the open they are quite visible and create a potential hazard to people walking about since it is possible to trip over the wires if they are low, or run into them if they are a bit higher in the air.

Some broadcast stations make use of a *counterpoise screen* which consists of many wires laying on, or buried a few inches in the ground. The wires fan out from the base of the antenna and are usually about 0.3 wavelength long. As many as 200 wires may be used in the screen. The screen serves to conduct the ground current out from the antenna base into the soil thus providing a large, relatively low resistance ground area beneath the antenna. This device is not the same as a tuned radial wire since the ground screen is not resonant. A few buried radial wires do not duplicate the screen since burying the radial detunes it to such an extent that the resonant concept is lost.

If you have the space to bury a large number of counterpoise wires a few inches below ground level you can make an acceptable substitute for the resonant radial wires. In most cases, however, it is easier to hide a wire or two than it is to construct an effective counterpoise screen.

BAND	RADIAL GROUND WIRE LENGTH	
	FEET	METERS
160 LOW	123′ 0″	38.5
160 HIGH	120′ 0″	36.6
80	63′ 0″	19.3
40	32′ 6″	10.0
30	23′ 0″	7.0
20	16′ 6″	5.1
17	12′ 9″	3.9
15	11′ 0″	3.3
12	9′ 3″	2.8
10	8′ 3″	2.0
6	4′ 6″	1.35

Fig. 4 TUNED RADIAL WIRE is cut to approximate quarter wavelength for each band. Flexible, insulated #14 stranded wire is suggested for general use. End of radial wire must be taped as it is ''hot'' and can cause a nasty r-f burn if it is touched when transmitter is operating. Radial wire should run close to (but not touching) the ground, or may run along floor or baseboard of the operating room.

An End-Fed Single Wire Antenna Tuner

One of the secrets of the end-fed antenna is the use of an antenna tuner in conjunction with an SWR meter. An imported SWR meter that will do the job may be purchased for about ten dollars, so it is unecononomical in most cases to construct one. The antenna tuner, however, must be constructed by the operator. The tuner is not complicated and can be built on a piece of plywood for simplicity. The schematic of a practical tuner is shown in Figure 5 and the layout is shown in Figure 6. It consists of a tapped air-wound coil and a tuning capacitor. The coil and capacitor are arranged so the connections may be changed to suit the electrical requirements of the antenna.

Tuner Construction

Placement of the tuner components is shown in the photograph. The unit is built on a base of 1/2-inch thick plywood 9'' long and 8'' deep. Rubber feet are placed at the underside corners to prevent it from scratching the surface upon which it sits. The panel is a piece of hard surface *Masonite* 9'' long and 6'' high. Panel and base are nailed and glued together.

The tuning capacitor is mounted to the base on two 1/2-inch ceramic standoff insulators. Directly behind the capacitor is the coil supported on a thin plastic sheet 8'' long and about 2'' wide. The plate is mounted on two 2'' insulators bolted to the base. The coaxial input receptacle is bolted to a small aluminum mounting bracket

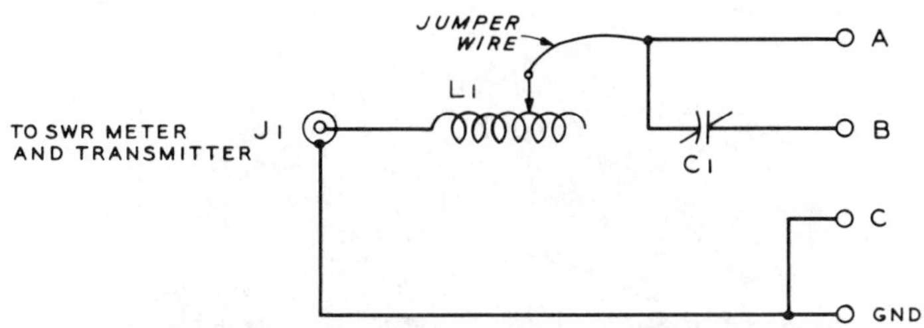

Fig. 5 SIMPLE TUNER FOR END-FED ANTENNAS requires only coil and tuning capacitor. The coil is 38 turns of #16 tinned wire, 6 turns per inch, 2" diameter (I-Core Air-Dux 1606 or equivalent). Tuning capacitor is 350 pF, 2 kV working voltage (Johnson 154-2 or equivalent). Tuning dial has metal plate insulated from condenser shaft to reduce hand capacity (Waldom 292 or equivalent). The coaxial receptacle is an Amphenol 83-1J to match common PL-259 coaxial plug. Various insulators are ceramic(Johnson 135 series or equivalent).

* * *

screwed to the left end of the base and terminals A, B, C and GND are 1" insulators bolted to the right end of the base.

Connections between the components are made with 1/8-inch wide strap cut from copper flashing material. The center pin of the coaxial receptacle is wired to the nearest end of the coil and the receptacle frame is connected to the GND terminal. A flexible jumper of #12 insulated, stranded wire about 6" long is used for the adjustable connection from terminal A to the coil. The end of the jumper is soldered to a copper "alligator" clip. Care should be taken in fastening the clip to the coil so that it does not touch an adjacent turn. This can be prevented by depressing every other turn of the coil along one side with the blade of a screwdriver.

A short external wire jumper is placed between terminals B and C in most instances and the antenna is attached to terminal A. The radial ground wires and external ground (if any) are attached to terminal GND. The tuner now functions as an L-network. By removing the jumper and transferring the antenna from terminal A to terminal B (leaving terminal C blank) the unit functions as a series tuned network.

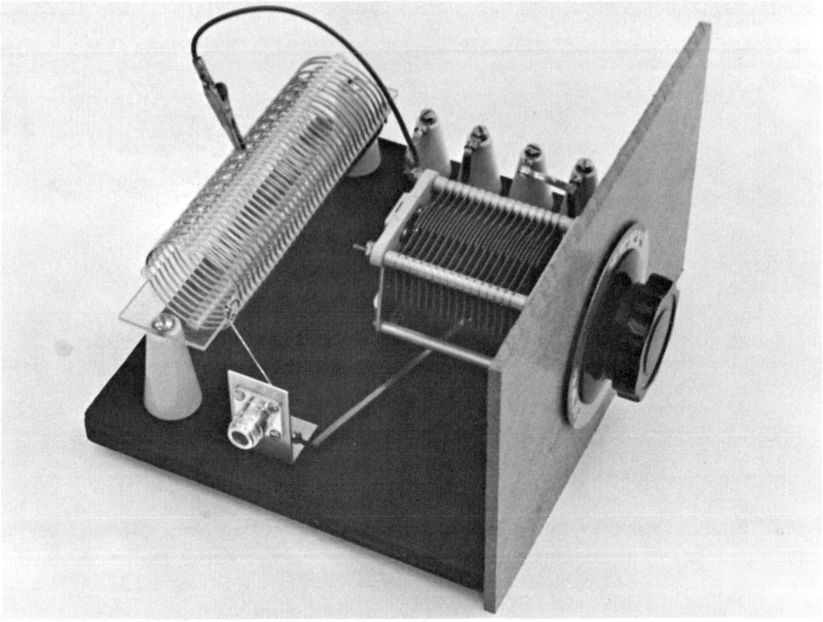

Fig. 6 SIMPLE ANTENNA TUNER for end-fed wire makes all-band operation simple. Tuner is built on plywood base with masonite panel. Coaxial receptacle for connection to transmitting equipment is at left with tapped coil directly behind the tuning capacitor. Every other turn of coil is depressed along one side so that easy connection may be made to it by a copper alligator clip. Antenna connections are along right edge of base. For majority of antennas the two center terminals are strapped together, with antenna connecting to the rear terminal and the ground connecting to the front terminal (see text).

Generally speaking, except for very short antennas, the first set of connections will be used for the majority of antenna lengths. The second set will apply for short antennas and whips.

Remember that the proper radial should be used for each band to establish a good radio ground for your antenna system, even though you may be able to tune everything up without the radial, or with the wrong length radial in the circuit. Use of no radial, or the incorrect one, is a direct invitation for TVI and erratic transmitter operation .

The End-Fed Single Wire Antenna

Now that you have built the tuner, how about an antenna? Any random length of wire will work and you can actually load up your gutter pipes, the bed springs or any handy, metallic object insulated from ground -- although this is not recommended!

The best antenna, of course, is straight, high and in the clear. An ideal end-fed antenna would be about 110 to 120 feet long for operation on all bands from 160 through 10 meters. For operation on 80 meters and higher frequencies, the antenna can be reduced in length to about 65 feet. Operation on 20 meters and higher can be accomplished with a 45 foot long antenna.

It is emphasized, however, that the tuner will work with *any* length antenna on *any* band as long as the antenna is at least 1/10 wavelength long on the lowest frequency band. Length of the end-fed antenna is *not* critical and the best suggestion is that the antenna be long enough to reach from the tuner to the far point of attachment! Keep it as high as possible and as much in a straight line as possible and it will work well for you.

Using the Tuner and the End-Fed Antenna

A single wire antenna from 60 to 100 feet long is a good one to start with to learn to adjust the tuner. The tuner will work with shorter lengths (down to 20 feet or so) or longer lengths but tuner adjustment becomes more critical as the antenna becomes shorter.

For preliminary adjustment, the antenna is connected to terminal A, terminals B and C are connected together and a radial ground wire for the band in use, is attached to the *ground* terminal of the tuner. If a good external ground is available, it should also be connected to terminal GND. The 50 ohm coaxial line from the transmitting equipment is connected to an SWR meter and from there to receptacle J_1 on the tuner. Using carrier injection on SSB, or other means, enough rf power is fed to the antenna system to cause a full scale deflection on the SWR meter when it is set to read *forward* power. A few watts will do the job. Don't apply full power during tune up, or you might burn out the SWR meter. Next, switch the SWR meter to read *reverse* power and adjust the tap on the tuner coil and the setting of the tuner capacitor until a null (lowest reading) is obtained on the reverse SWR reading. Tap along the coil a few turns at a time and rotate the capacitor through its full range on each test. A tap point should be found at which reverse SWR reading will drop to near-zero upon adjustment of the capacitor. Further experimentation will show that the number of turns in the coil and the setting of the capacitor are somewhat interlocking and the exact number of turns used is not very critical, as the setting of the capacitor can be varied to compensate for too many or too few turns.

If, by chance, the reverse SWR reading cannot be made to drop, the antenna tuning unit should be reconnected. The jumper between terminals B and C is removed and the antenna connected to terminal B. Adjustment of the coil tap and capacitor setting are then made so as to realize the near-zero reverse SWR reading on the meter. Once the proper settings are found, they should be recorded in your log book for future use, or typed and pasted to the tuner panel.

The 3/8 Wavelength Two-Band Marconi Antenna

Some end-fed antennas require no antenna tuner, or else a very simple one consisting only of a variable capacitor. One of the latter types is the 3/8 wavelength long Marconi antenna which may be operated both on the fundamental frequency for which it is cut and also on the second harmonic frequency, thus providing good two-band performance. For example, the antenna may be cut for the middle of the 80 meter band to provide good performance over the whole band. It will then operate over the whole 40 meter band on its second harmonic. Such an antenna is shown in Figure 7 with dimensions given for 160/80, 80/40 and 40/20 meter operation. On the lower of the two bands, the antenna acts as a series-tuned 3/8 wavelength Marconi and on the higher band the antenna acts as a self-resonant 3/4 wavelength Marconi. Operation on the higher band is determined by choice of antenna length, the antenna then being resonated to the lower band by means of series capacitor.

As with any end-fed antenna, radial ground wires are required, at least one wire for each band of operation. Radial lengths may be determined from Figure 4, given earlier in this chapter.

The tuning capacitor is shorted out of the circuit for high band operation, the antenna being connected directly to the coaxial transmission line (or directly to the transmitter) through an SWR meter. The exact resonant frequency of the antenna may be determined by disconnecting the antenna from the line or SWR meter and connecting it directly to ground via a two turn loop of hookup wire, about one inch in diameter. The loop is coupled to a grid-dip oscillator and the frequency is checked by noting the meter dip, while monitoring the oscillator on a nearby, calibrated receiver. If the antenna is resonant at too low a frequency, it may be shortened a few inches at a time until a new resonance point is established. Likewise, lengthening the antenna will lower the resonant frequency.

The capacitor is cut in the circuit on the low band and adjusted

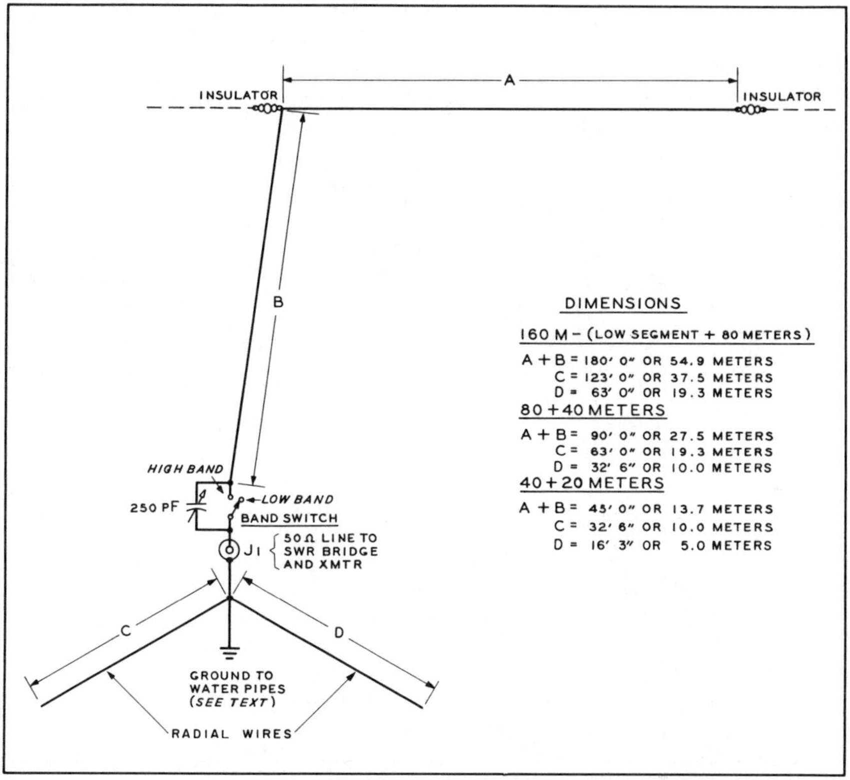

Fig. 7 TWO-BAND MARCONI ANTENNA functions as 3/4 wavelength self-resonant antenna on high band and as a tuned 3/8 wavelength antenna on low band. Low band resonance is established by adjustment of tuning capacitor. High band resonance is established by trimming length of A section. Capacitor is 250 pF, 2 kV (Johnson 154-1 or equivalent). Separate radial ground wires are required for each band .

until the lowest possible value of SWR is indicated on the SWR meter. No adjustment to antenna length need by made for low band operation.

The Twin-Lead Marconi for 80 or 160 Meters

Operation on the two lowest frequency ham bands poses a problem because of the length of the antenna required and because of the radiation resistance of the antenna, both of which are quite low. The typical ham antenna for these bands is very close to the earth in terms of wavelength and this brings about these difficulties.

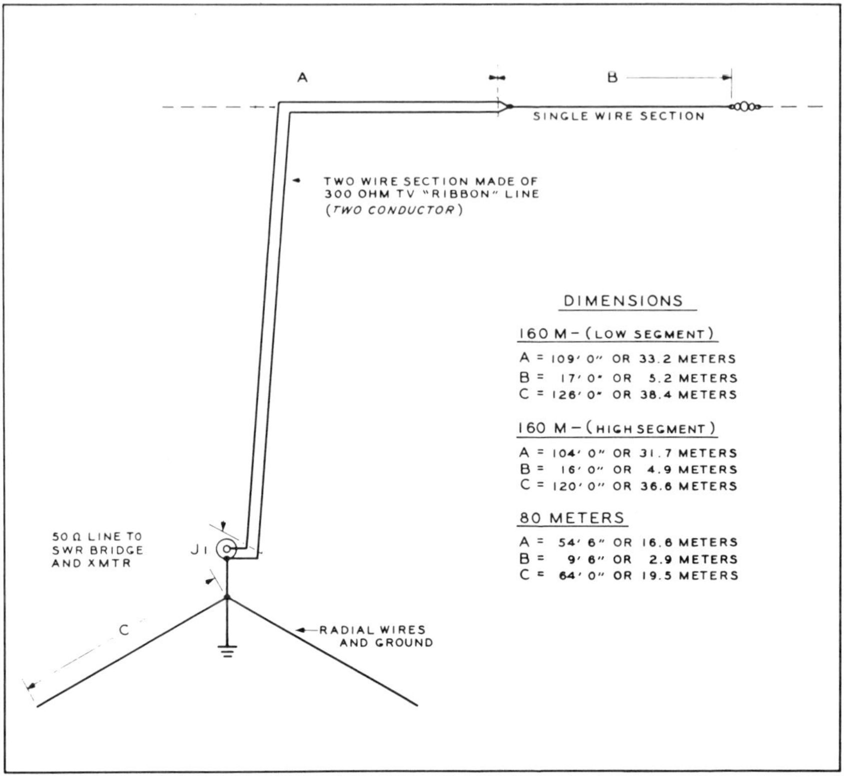

Fig. 8 TWIN-LEAD MARCONI provides efficient operation on one amateur band. Impedance match achieved by two-wire portion of flat-top raises impedance of antenna to match 50 ohm transmission line. One or two radial ground wires and good ground connection are used for best results.

The folded antenna is made of a length of 300 ohm "TV ribbon line" connected to a short length of wire, as shown in Figure 8. No antenna tuner is required with this antenna. The conductors at the far end of the "ribbon"are shorted together and connected to the additional length of wire. One of the wires of the twin lead is grounded at the transmitter end of the antenna and the other is attached to the antenna terminal of the transmitter, or is connected to it via a length of 50 ohm coaxial cable. The "ribbon" forms a kind of linear impedance matching transformer which is very efficient and practical.

Chapter 9

Vertical Antennas You Can Build

(No Space for an Antenna? Go Straight Up!)

The vertical dipole antenna was discussed in Chapter 4. It's an effective DX antenna which radiates, or sprays, radio energy in all directions around it. A version of the vertical dipole is the popular *ground plane* (G-P) antenna, shown in Figure 1. The ground plane antenna consists of a vertical antenna section, or *radiator,* mounted above several horizontal rods, or *radials*. The radiator and radials are all usually about a quarter-wavelength long. The length of a quarter-wave element is:

$$\text{Length (feet)} = \frac{234}{\text{Frequency (MHz)}}$$

The vertical section is considered to be the antenna proper, and the radials establish an artificial radio ground, or ground plane (hence the name) at the base of the radiator. Looking at it another way, the ground plane antenna is a vertical half-wave dipole, fed from below, with the bottom half of the dipole split into separate radials which are swung up into the horizontal plane. The antenna is fed at the junction of the vertical section and the radials with a coaxial transmission line.

A general coverage antenna such as this is very effective when it is desired to communicate with stations located in many different directions about you without the necessity of turning a beam antenna in the direction you wish to work. The ground plane antenna provides world-wide coverage for DX on the high frequency bands and general purpose coverage on the VHF bands for working mobiles that move about the countryside.

Signals from a ground plane antenna spread out much as the concentric circles spread out from a stone cast into a quiet pond. On the other hand since the ground plane radiates and receives in all directions, it

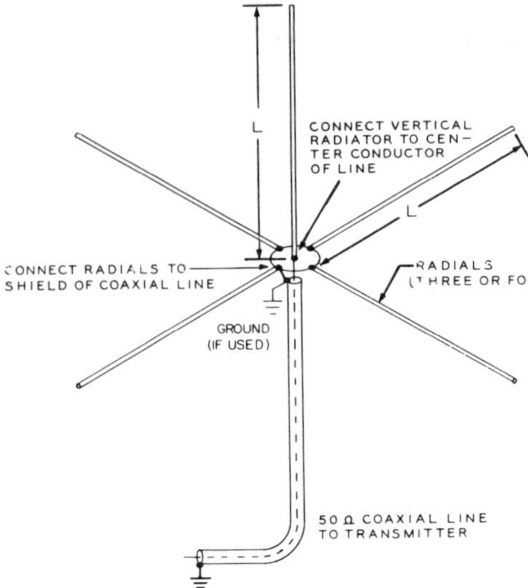

LENGTH (L) OF RADIATOR AND RADIALS

BAND	FEET	METERS
80	63′ 0″	19.3
40	32′ 6″	10.0
30	23′ 0″	7.0
20	16′ 6″	5.1
17	12′ 9″	3.9
15	11′ 0″	3.3
12	9′ 3″	2.8
10	8′ 3″	2.0
6	4′ 6″	1.4

Fig. 1 GROUND PLANE is effective DX antenna for H-F bands. Radials should be sloped down at angle from horizontal for best SWR and may be used as guys.

does not provide the user with any protection against interfering signals coming in from a different direction -- as does a beam antenna -- since all signals on one frequency are heard, regardless of the incoming direction.

The ground plane is a compact DX antenna that is well regarded on the amateur bands. Its main fault is that it is very responsive to man-made noise (ignition noise, line noise, etc.) which is mainly vertically polarized. The horizontal dipole, on the other hand, shows some discrimination against such noise. If you live in an electrically noisy location, the use of a vertical antenna, particularly on the lower frequency bands where noise is severe, should be tempered with caution.

Build a High Frequency Ground Plane Antenna

A ground plane antenna for a high frequency amateur band is a worthwhile and relatively simple project. A ground plane antenna for 80 meters is quite big, but ground planes for the higher bands are fairly compact and kits are available for 10, 15, or 20 meter ground planes at modest prices. You can build your ground plane out of readily available parts for less cost than an equivalent "store-boughten" model, if you wish.

Here are the plans for a simple G-P antenna that will do a good job for you. A typical installation is shown in Figure 2. The antenna consists of a vertical whip made up of one or more sections of aluminum tubing affixed to a wooden mast or other support. Three or four semi-horizontal radials are used, which double in duty as guy wires for the mast.

The ground plane is a single band antenna, and all dimensions are calculated for proper operation across the entire band. Tubing diameter of the vertical whip section is not critical, and diameters of 1/2 inch to 1-1/4 inches may be used. Commercial aluminum alloy tubing of 6063 or 6061 grade is recommended as a good compromise between strength and ability to resist corrosion. If more than one section of tubing is required, telescoping sections should be used to make a neat and ship-shape joint that will withstand weather and wind.

To make this connection properly, one end of the larger tube is cut through both walls, on a line parallel with the center axis of the tube. All burrs are carefully removed from the wall of the tube after the cut. The matching sections of the two tubes should be sanded bright and cleaned to improve electrical conductivity and to lessen the possibility of seizure after the tubes are telescoped.

Before the element is assembled, steps are taken to prevent corrosion at the joint. A special antioxidizing compound is smeared lightly over the mating pieces of tubing *(Penetrox A,* manufactured by Burndy Co., Norwalk, Conn.). After the tubes are telescoped, this compound forms an air-tight seal, preventing corrosion. The compound is a good electrical conductor and provides a low resistance, trouble-free joint. A 5 oz. tube of *Penetrox* is sufficient to coat many antenna joints.

The last step is to clamp the joint securely with a tubing or hose clamp to prevent movement between the tubes in a wind. If the tubing sections do not make a close fit, the joint can be shimmed with *thin* strips of aluminum.

Assembly of the Ground Plane Antenna

The aluminum whip section is bolted to the wooden mast or support by means of two U-bolts as shown in the illustration. Since the r f voltage at this point is very low, it is not necessary to employ any insulation other than the wood of the mast. The mast, by the way, should be given two coats of outdoor paint to protect it from the weather.

At least three radial wires are required to guy and support the mast. The wires are cut to length, allowing about three extra inches at each

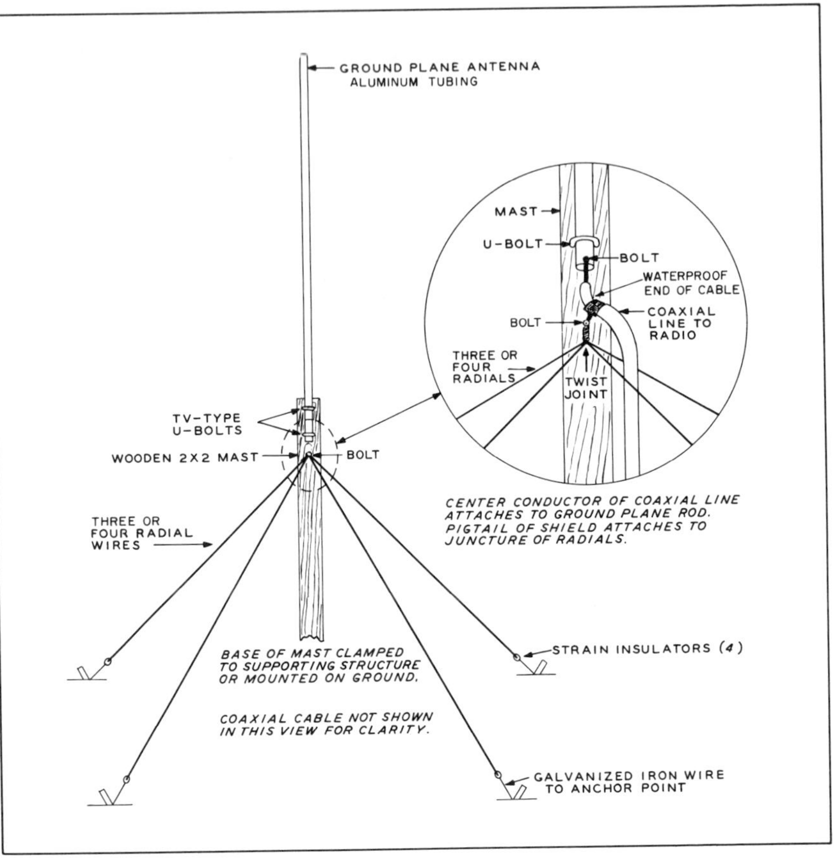

Fig. 2 INEXPENSIVE GROUND PLANE is bolted to wood mast with TV hardware and guyed in position by sloping radial wires. Antenna can be roof or ground mounted and is fed with 50 ohm coaxial line.

end. The far end of each wire is passed three inches through the eye of an egg insulator (strain insulator) and then looped back, twisted upon itself and soldered. At the mast, the free ends of the radials are twisted together for about two inches and soldered to form a single, heavy lead. This lead is firmly twisted about a through-bolt in the wood mast placed an inch or so below the bottom of the vertical aluminum tube. Overall radial length is measured from the tip of the wire to the bolt. Wires or ropes may be tied to the free eye of each insulator at the tips of the radials, making the radials serve as supporting guy wires for the wooden mast.

The feedline is prepared as described in Chapter 5. The center conductor is soldered to a lug which is bolted to the vertical whip section and the radials are attached to the pigtail braid made from the outer shield. The joint is sealed with the *General Electric* RTV-102 "goop" mentioned in the previous chapter.

The antenna and mast are erected in a vertical position and the radial-guy wires fanned out below the structure. Mast height and guy wire length can be adjusted so the radial wires drop down at about a 45 degree angle (not critical) from a horizontal plane. The transmission line is led down the wooden mast until it clears the radials after which it may be led away in a horizontal direction to the station. The antenna is now ready for use -- and good DX reports!

The Sloping Radials

The sloping radials serve two purposes in this antenna. First, they act as guy wires. Second, they act as a simple matching transformer which provides a near-perfect match between the ground plane antenna and the 50 ohm coaxial line. The radiation resistance of a true ground plane antenna (one having horizontal radials) is about 30 ohms. The lowest value of SWR that may be achieved in this case is the ratio of line impedance to the antenna radiation resistance. This ratio is 50/30, or an SWR of 1.66. By drooping the radials about 45 degrees, the radiation resistance of the modified ground plane will be very close to 50 ohms, accurately matching the characteristics of the transmission line and reducing the SWR to less than 1.2 or so. Thus, maximum power is transferred from transmitter to antenna, and radiated into space.

The Multi-Band Ground Plane Antenna

The vertical ground plane antenna provides a low angle of radiation that makes it a good DX antenna for any amateur band. Placing two ground planes in one back yard is a complicated task, as the amateur wishing two band operation is liable to hang himself as he moves about in the snarl of radial wires and feeders! Once again, the multi-band principle used so effectively with the dipole comes to the rescue, permitting two or more ground plane elements to be fed with a single feeder and taking advantage of a simple radial system made up of multiconductor cable. Here are the details of this compact and efficient antenna system.

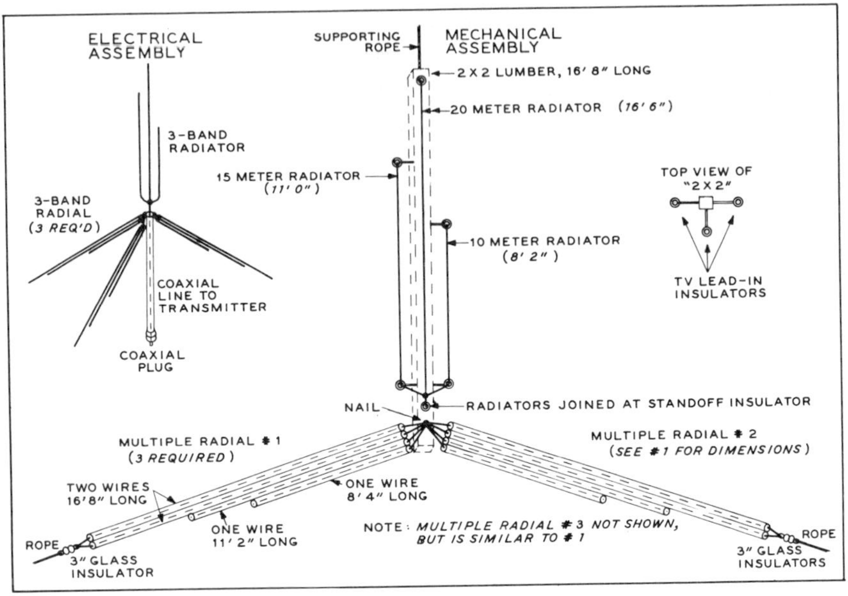

Fig. 3 TRI-BAND GROUND PLANE for 20, 15 and 10 meters is composed of three vertical antennas connected in parallel at the feed point and three radials made of multiple conductor TV cable. The antenna is mounted to a wood mast which may be supported from the top as shown, or bolted to a suitable support at the bottom. Radials are slanted down about 45 degrees from horizontal.

The DX Ground Plane for 20, 15, and 10 Meters

Shown in Figure 3 is a multi-band ground plane for the popular DX bands. It costs little to build, is easy to assemble, and simple to get up in the air if a handy tree or rooftop is nearby. The antenna is composed of three vertical 1/4-wave radiators fed in parallel at the base, working in conjunction with three multi-band radials which extend below the radiators in a slanting position. The junction of the radials and the radiators is fed with a 50 ohm coaxial line.

The expensive way to construct this antenna is to buy three pieces of aluminum tubing to make the vertical radiators. A far simpler and less expensive idea is to make the radiators of No. 12 copper wire strung along a sixteen foot length of "two-by-two" lumber. It is possible, moreover, to attach a rope to the top of the wooden support, toss the rope over a high branch in a tree and pull the ground plane into operating position high in the air without lifting your foot from the ground!

Antenna Construction

The antenna support is a sixteen foot length of 2" x 2" lumber (which usually measures about 1-1/2" x 1-1/2" on a side). Choose a straight, dry piece and give it two coats of outdoor varnish before you start to work. The radiator wires are cut to length, allowing a few inches on each end, and are fixed in position on the wooden support by means of 7½" wood-screw type TV lead-in insulators. The insulators are run into the support as shown in the illustration and the radiator wires are passed through the polyethylene inserts. Tie the wire to the insulator with cord at each termination. The bottom ends of the radiators are cleaned of insulation, twisted together and finally fastened to a small porcelain stand-off insulator screwed to the base of the pole. If you are lucky, you can pick out a "sixteen foot" length of lumber that actually measures about 16'8" long: long enough to take the full length of the 20 meter radiator wire. A shorter piece of lumber requires that the top TV lead-in insulator be run into the top of the support, with the 20 meter wire gracefully making a curve over the top of the pole. This will in no way hinder the operation of the antenna. You can make a little hook in the top end of each radiator wire and wrap the wire with vinyl tape just above the insulator to hold it securely in place.

Two or three radials are required, each radial made of inexpensive four conductor TV rotator control cable. The radials run down from the base of the radiators towards the ground at an angle of approximately 45 degrees. In addition to their electrical function they also act as guy wires to steady the bottom end of the wooden support. As only three wires are required in each radial assembly, two of the four conductor wires are connected in parallel at each end to form the 20 meter radial, since these have the greatest physical strain upon them. The three radial sets are soldered together at the antenna end and anchored to a nail driven in the wooden pole just below the terminating insulator for the radiators.

The next step is to determine the length of coaxial line required to reach from the base of the antenna assembly to your shack. You can use the less-costly RG-58/U coaxial cable for transmitter power under 500 watts, and the larger RG-8/U coaxial cable for higher power. Carefully strip the outer insulation from one end of the cable for about six inches taking care not to damage the tiny wires of the outer shield. Now, using a small nail, carefully unbraid the shield as far back as the outer insulation, twisting the wires into a single conductor once they are free of the center insulation. This braid pig tail will support the weight of the line. Finally, trim the insulating core from the center conductor for a distance of four inches and seal the joint firmly using the RTV-102 sealant described previously, and taping the joint, after it dries, with vinyl electrical tape. Solder the pig tail to the junction of the radials and solder the inner conductor of the line to the termination point of the vertical radiators and your antenna is completed.

Antenna Installation

If you have a handy tree, a flat roof, or a convenient chimney your troubles are almost over! It is possible to hang the antenna by its tip from the upper tree branches, like a gigantic plastic icicle on a Christmas tree. The support will dangle like a pendulum, stabilized at the bottom end by the three radial wires. To accomplish an installation of this type all that is required is a tree, some string, a sturdy rope, and a steady eye and strong arm. Tie a small rock to the string and (being observant of nearby windows, etc.) toss the string over the highest branches of the tree, letting the string settle in a crotch of the branches. Once the string is safely over the branch and the rock down again at ground level the heavier rope may be pulled over the branch with ease. With the top of the ground plane firmly tied to one end of the rope it may be hoisted to the uppermost branches with little effort. Care should be taken to guide the antenna through the tree branches on its ascent, protecting the wires of the assembly from being snarled or snagged during the trip. Jockey the antenna into a vertical position clear of leaves and branches and steady the base by means of the radials. Slope the three radials down at some angle between 30 and 45 degrees to the horizontal and tie them off to nearby branches and your job is done.

If you have a flat roof on your house or garage a slightly different installation technique should be used. The bottom of the support may be spliced to a "christmas tree stand" type of mounting with the base of the antenna supported about six feet or so above the roof, guyed in place by the radials. Television mounts, clamps and other antenna accessories will come in handy, especially if you plan to fasten the antenna to a chimney. With a little ingenuity you can also fasten the assembly to a vent pipe on the roof of your house, or to a sturdy metal gutter pipe. Keep the base of the antenna high enough so that you can slope the radials downwards from the bottom of the radiator as shown in Figure 3. The slope angle is not particularly critical but provides a somewhat better value of SWR on the line than if the radials were run out in the horizontal plane.

Bonus Operation on 40 or 80 Meters

Yes, it is possible to operate this antenna system on either 40 or 80 meters provided you choose the right length of coaxial line to span the distance between antenna and transmitter! It is merely necessary to connect the outer conductor of the coax to the inner conductor at the transmitter end of the line, connecting the whole "shooting works" to the output terminal (inner conductor of the coaxial socket) on your transmitter. Run a short, heavy wire from the chassis of the transmitter to a nearby water pipe and attach a quarter-wavelength radial ground wire to the equipment, as described in the previous chapter. The antenna and coaxial line will now act as an end-fed quarter wave Marconi antenna with the shield of the coaxial line doing the radiating and the ground plane antenna merely contributing a bit of top loading effect. For 40 meter operation the coaxial line should be about 50 feet long, and for 80 meter operation it should be approximately 106 feet in length. There is nothing mystical in these dimensions and you might find that some intermediate length will work well on both bands for you. You can "zero in" on the correct length by observing the SWR reading at the station and trimming the coaxial line length for minimum SWR.

Build a Cobra Vertical for 10 or 6 Meters

Are you interested in a simple, inexpensive vertical dipole antenna that can be built in an hour or so, and provides good performance on 10 or 6 meters? Useful for emergencies and portable work, the Cobra antenna was named by an enthusiastic user who saw in the rf choke coil and vertical section a resemblance to the weaving reptile of the snake charmer! What an imagination!

The Cobra antenna is a vertical half-wave dipole composed of an upper quarter-wavelength section made of copper wire and a lower quarter-wavelength section made up of the braided outer conductor of the coaxial line (Figure 4). The simple antenna is suspended at the top from an insulator and a length of rope and hung from a tree or other handy support. The Cobra is fed at the base by a coaxial line, the end of which serves as the bottom portion of the antenna. The remainder of the transmission line is isolated from the antenna portion by a home made choke coil made of a length of the line wound around an inexpensive ferrite core.

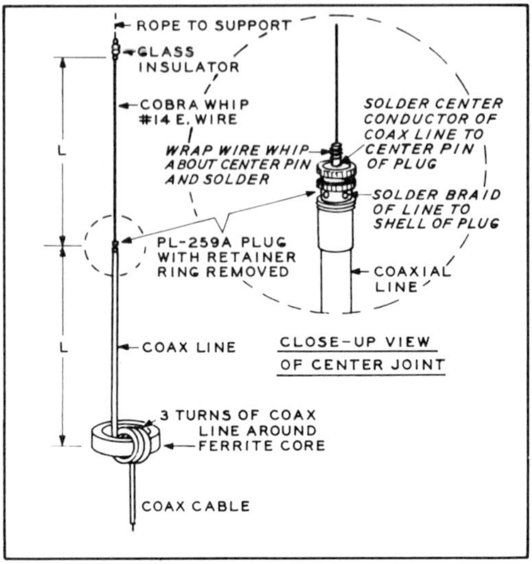

Fig. 4 COBRA ANTENNA employs section of coax feedline as lower part of antenna. Length L is given in Fig. 1. Ferrite core is #CF-123. Write for price and information to: Indiana General Corp., Crow Mills Rd., Keasby, N.J. 08832. Core is Q-1 material, 2.4" diameter and about 0.5 inches thick.

Antenna Assembly

Your first job is to construct the lower antenna section out of a length of RG-58A/U coaxial line. A PL-259 style plug and UG-175/U reduction adapter are placed on the end of the line, but the threaded retainer ring of the plug is *not* used. The plug merely serves as a convenient terminating device for the end of the cable. Trim the line carefully, as explained earlier in this Handbook, and solder the outer braid of the cable to the shell of the plug through the four small solder holes in the plug. Use a very hot iron or gun with a small tip for this job. Solder the center wire of the cable to the center pin of the plug (after your review Chapter 5). Next, the ferrite core is placed on the line. A three turn coil is made about the core by passing the line through the core three times. The coil should be about four inches in diameter. Tape the turns together and firmly to the core, which is quite fragile. Now, cut a length of copper wire for the upper sections of the antenna. Clean and solder one end of the wire to the center pin of the coaxial plug, wrapping the wire several times around the pin to make a strong mechanical joint. Attach the top insulator to the wire so that the distance between the tip of the wire and coaxial plug is correct. Wrap the wire securely back upon itself. The last step is to coat the coaxial plug center joint with *General Electric RTV-102* waterproofing compound and wrap it with vinyl tape.

Antenna Installation

When you are finished the Cobra is ready to work. It can be hoisted into a nearby tree with the aid of a string and a stone. Tie a light string to the stone and (watching out for nearby windows and spectators) toss the stone over the highest branches of the tree. A husky ham can usually make a sixty foot toss with ease. (How muscular are you?) Once the string is safely over the branches and the rock is down again to ground level, a heavier rope may be pulled over the branches. The top insulator of the Cobra antenna is attached to the rope and the antenna hoisted up as far as it will go, swinging in a vertical position like a gigantic icicle on a Christmas Tree! The Cobra operates well over the 10 and 6 meter amateur bands with an SWR reading of between 1.3 and 1.8 and is a handy, compact emergency antenna to carry in your automobile at all times.

The 5/8 Wavelength DX Antenna for 20, 15 or 10 Meters

Are you interested in a simple vertical antenna that gives you a 3 decibel power gain over a ground plane? If so, this is the antenna for you. Shown in Figure 5, the extended 5/8 wavelength vertical for 20 meters consists of a 41'6'' length of telescoping sections of aluminum tubing, fed at the base by a simple coaxial matching transformer. Two or three 16'6'' radial wires are used, and these are laid out from the base of the vertical antenna in random position, running horizontally away from the antenna.

The matching transformer is made of two lengths of 50 ohm line. The small RG-58/U may be used for powers up to 500 watts, or less. RG-8/U line is used for a high power installation. The line sections are coupled to the main transmission line by means of a coaxial "T-adapter" (*Amphenol* 83-1T).

The coaxial line from the T-adapter to the base of the antenna for 20 meter operation is 8'8'' long, including the PL-259 plug. The short circuited line section is 2'1'' long, including the plug. Together, the two lines form an electrical quarter-wavelength transformer, with the main transmission line to the station tapped on the 50 ohm point.

The 5/8 wavelength vertical antenna is insulated at the base and the center conductor of the simple coaxial transformer is attached to the base, with the transformer shield attached to the radial wires.

If the measurements are followed, no adjustment need be made to the antenna system to provide good operation across the 20 meter band.

DIMENSIONS OF 5/8 λ ANTENNA								
BAND	L		R		A		B	
	FEET	METERS	FEET	METERS	FEET	METERS	FEET	METERS
20	41′ 6″	12.65	16′ 6″	5.31	8′ 8″	2.63	2′ 1″	0.64
15	27′ 9′	8.47	11′ 0″	3.34	5′ 10″	1.78	1′ 6″	0.46
10	21′ 3″	6.50	8′ 3″	2.52	4′ 4″	1.31	1′ ½″	0.31
6	11′ 8″	3.55	4′ 6″	1.37	2′ 5″	0.74	8″	0.20

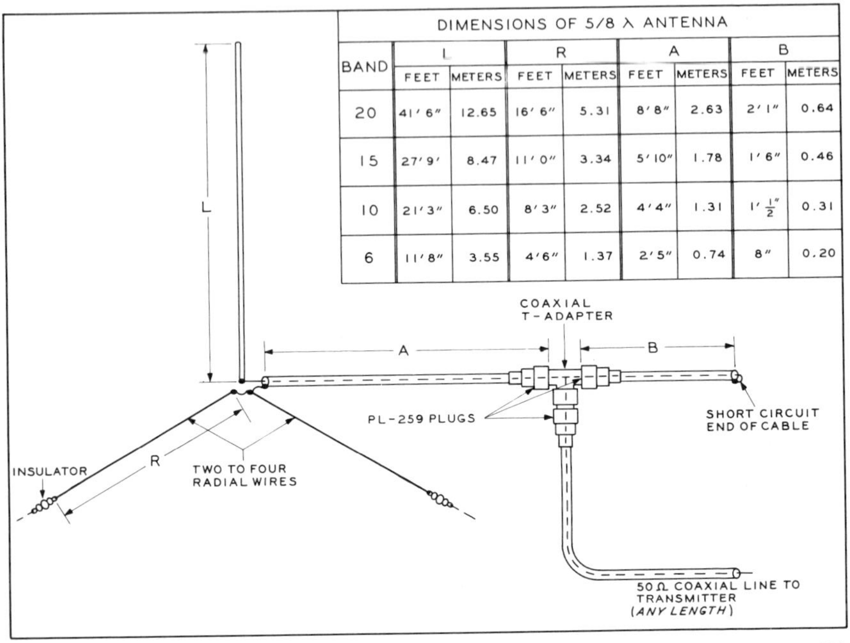

Fig. 5 SIMPLE COAXIAL TRANSFORMER provides impedance step-up between 50 ohm line and base of 5/8 wavelength vertical antenna. Coaxial line lengths do not include length of T-adapter (Amphenol 83-1T). RG-8/U type cable should be used for the transformer and connection to antenna should be carefully sealed to prevent moisture from entering line. Wrap T-adapter with vinyl tape and seal also, as it is not waterproof.

The SWR may be adjusted by varying the length of the antenna a few inches, one way or the other.

Dimensions for making a 5/8 wavelength antenna for the 20, 15, 10 and 6 meter bands are given in the illustration.

A Simple, Multi-Band Vertical DX Antenna

This versatile ground plane antenna designed by G3RFG of England may be used for operation on all the high frequency amateur bands between 160 and 10 meters. It's an extremely attractive installation for the amateur with a lean purse, living in an area that frowns upon structures more exotic looking than an unobtrusive vertical mast.

The complete antenna assembly is shown in Figure 6. It consists of a vertical radiator (A-B) attached to a short support post. The antenna itself is broken into two sections and a jumper (C) or loading coil (D) connects the sections, depending upon the band in use. On 10

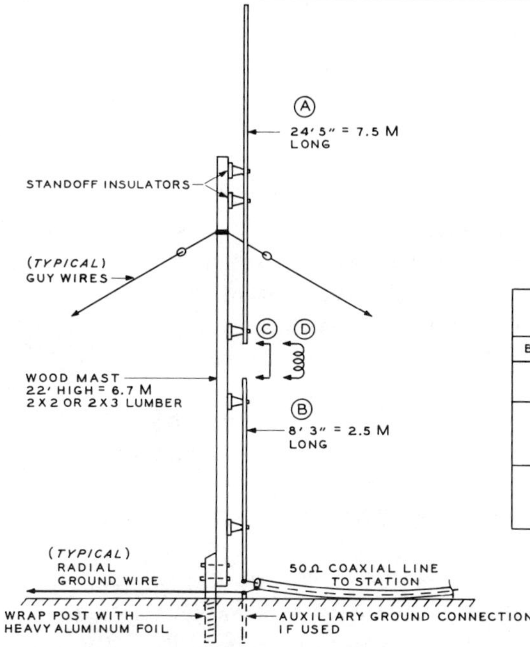

LOADING COIL (D)	
BAND	22 TURNS N° 14 ENAMEL
20	2" DIA., $2\frac{1}{2}$" LONG = 5 CM DIA., 6.3 CM LONG
80	30 TURNS N° 18 ENAMEL 2" DIA., 3" LONG = 5 CM DIA., 7.5 CM LONG
160	55 TURNS N° 18 ENAMEL 2" DIA., 5" LONG = 5 CM DIA., 12.6 CM LONG

Fig. 6 ALL H-F BANDS may be operated with this versatile antenna. Loading coil is used for 20, 80 and 160 meter bands, and jumper (C) is used for 15 and 40 meter bands. A separate radial ground wire is required for each band. Space between antenna sections is 6 inches. The antenna may be mounted on the roof or directly at ground level. If a ground post is used, it should be wrapped with heavy aluminum foil to prevent rot. Auxiliary ground rod driven four to six feet into the soil in addition to radial wire is recommended for 160 and 80 meter operation.

meters the gap between the sections is left open and the antenna acts as a 1/4 wavelength ground plane.

With the jumper in place, the antenna acts as a 40 meter ground plane and also as a 15 meter, 3/4 wavelength extended vertical antenna. A loading coil modifies the antenna into a 3/4 wavelength antenna for 20 meter operation or a loaded vertical for 80 and 160 meter work. The loading coil is placed so that it may easily be reached from the ground, yet high enough so that it will not become an attractive plaything for neighborhood children.

The antenna uses a set of radial ground wires to form an efficient electrical ground. An auxiliary ground connection, if available (as discussed in Chapter 8), should be used along with the radials.

All in all, this interesting antenna is just about the best combination of economy and efficiency that can be found and its use is recommended for multi-band operation in restricted areas.

Antenna Construction

The antenna is made up of 12 foot lengths of 1'' diameter aluminum tubing. A shorter section is cut from one length to make the bottom section. The two remaining sections, plus a short piece cut from the first section, are then joined to make the top section. Short lengths of telescoping tubing are used to make joint splices, which are held firmly together with 6-32 nuts and bolts. After assembly, the sections are given a good coat of aluminum paint to protect the tubes and joints from corrosion.

The antenna sections are next clamped to a 22 foot wood mast, made of seasoned 2 x 2 lumber. Small ceramic standoff insulators are used to do this job, two being used for the bottom tube and three for the top tube. The break between the top and bottom sections is about 2 inches, as noted.

The connections at the break may be made with large banana jacks and plugs. The jacks are bolted to the tubes by means of thin aluminum strips, and matching plugs are affixed to the jumper and loading coil.

The radials for 10, 15 and 20 meters are cut from a length of multiple conductor TV rotor cable. Individual radial wires are used for 40, 80 and 160 meters.

Antenna Adjustment

The resonant frequency of the antenna on 10 meters is determined by the length of the bottom section of tubing and is relatively uncritical. With the jumper in place, resonant length for 40 meters may be set by adjusting the top section of the antenna. This automatically adjusts 15 meter operation at the same time, as the antenna works on its third harmonic on the higher frequency band. The resonant frequency on the 20, 80 and 160 meter bands is set by adjusting the number of turns on the loading coil.

The antenna tunes rather broadly for 10, 15, 20 and 40 meter operation and probably no adjustments need be made for these bands. As the antenna is quite short compared to a full-size ground plane on 80 and 160 meters, the loading coil should be adjusted with the aid of a grid dip oscillator for the portion of the band in which operation is contemplated. To accomplish this, the feedline is removed at the base of the antenna and the antenna connected directly to the radial ground wires through a small two turn link coil. The coil of the grid dip oscillator is coupled to the link and the resonant frequency of the antenna noted by monitoring the oscillator with a calibrated receiver. To lower the antenna frequency, additional turns should be added to the loading coil and to raise the antenna frequency, turns are removed from the coil.

A Simple, Build-it-Yourself 5/8 Wavelength Ground Plane Antenna for 2 Meters (Great for FM Operation!)

Do you need a simple and inexpensive whip antenna that gives you about 3 decibels gain over a quarter-wave ground plane? That's equivalent to *doubling* your transmitter power. This simple antenna is just the thing for 2 meter FM operators who are "on the go". While many FM transceivers come with a portable whip antenna, much better results may be achieved with the use of a full-size 5/8 wavelength antenna, mounted high and in the clear. It will enable you to "break" a repeater at an extreme distance and provide extended ground wave coverage as well.

This 5/8 wavelength ground plane antenna may be used with any 2 meter equipment, even the small handi-talkie units. It is designed to be rugged and light enough to be hauled up into a tree, mounted atop a light mast or atop an existing antenna, or used as a mobile antenna on your car or camper.

Antenna Assembly

This design makes use of a UHF-type coaxial plug (PL-259) as the base of the antenna (Figure 7). The whip antenna is made of a length of hard-drawn copper wire (or copper plated steel wire) and is fitted into a plastic cylinder mounted into the open end of the plug. The cylinder serves as a form upon which is wound the small matching coil which transforms the whip impedance to 50 ohms, matching it to the 50 ohm coaxial line used with most VHF equipment. The top end of the whip is bent into a small loop facilitate hoisting the antenna into a tree and to prevent the operator from putting his eye out when erecting the antenna.

For operation on a vehicle, the whip assembly is merely screwed into the receptacle of the body mount in place of the usual quarter-wave whip.

For use as a remote antenna to be mounted atop a mast, a set of four quarter wavelength radials are required. The simplest way to make

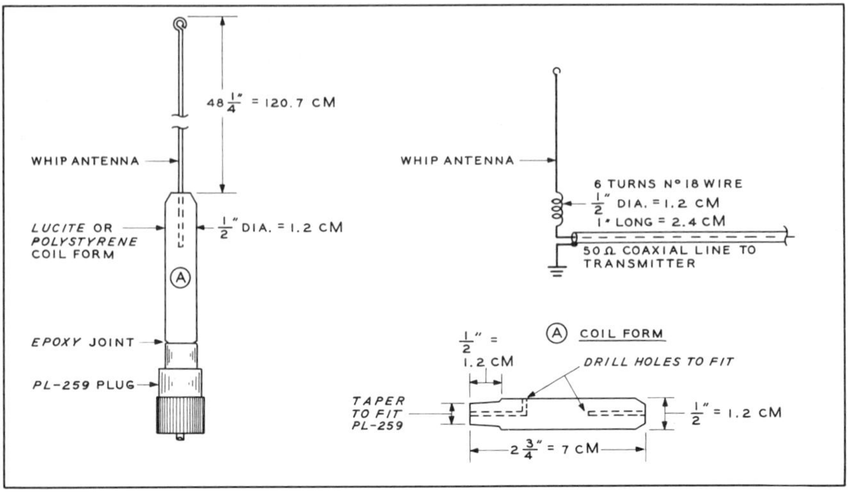

Fig. 7 EXTENDED 5/8 WAVELENGTH WHIP for 2 meter band provides improved coverage and is recommended for FM operation. Whip antenna can be made from replacement CB walkie-talkie antenna, or as described in text. Coil form is drilled at one end to accept whip and other end is drilled to pass wire connection from center pin of coaxial plug. The wire is "fished" through side hole and used to wind small 6 turn loading coil. For portable service, 19" radials should be used with the antenna.

these up is to solder four lengths of hard drawn copper wire or light copper tubing to the corners of an *Amphenol SO-239* receptacle. The antenna is screwed into the receptacle and the coaxial transmission line is soldered to the terminals of the receptacle. The antenna plus the modified receptacle form a complete 5/8 wavelength ground plane assembly ready for mounting on a mast or other support.

The Loading Coil

The 5/8 wavelength whip is matched to the transmission line by a small loading coil mounted at the base. The coil is wound on a 1/2-inch diameter form (A) cut from a plastic rod. One end of the form is drilled to accept the antenna and the other end is turned down in a lathe until it makes a force fit within the barrel of the PL-259 plug. Before the form is pressed into the plug, a short length of No. 16 tinned wire is soldered to the center pin of the plug and passed through the two small holes drilled in the form. The wire is brought out through the side of the form so that it may make contact with the base end of the coil.

The end of the form is coated with epoxy glue and pressed into the barrel of the plug. When the joint has hardened, the antenna is pressed into the hole in the opposite end of the form and epoxied into place. The last step is to wind the coil on the form, soldering one end to the base of the antenna and the other end to the connecting wire projecting through the side hole in the form. When completed, the coil should be given several thin coats of *Krylon* spray to waterproof it.

Antenna Adjustment

The antenna length is not critical and it may be tuned "on the nose" by altering the length a quarter-inch at a time while observing the reflected power on an SWR meter inserted in the transmission line. A temporary whip made of soft copper wire a few inches longer than normal can be used for this tuning process. The whip is shortened a fraction of an inch at a time with a pair of wire cutters. Once the point of minimum SWR is found, the final whip may be made of more rugged wire or light tubing.

If a more rugged antenna is desired, collapsible CB whips of various lengths may be purchased at a modest cost. The hole in the end of the form is drilled to fit the whip which is epoxied in place.

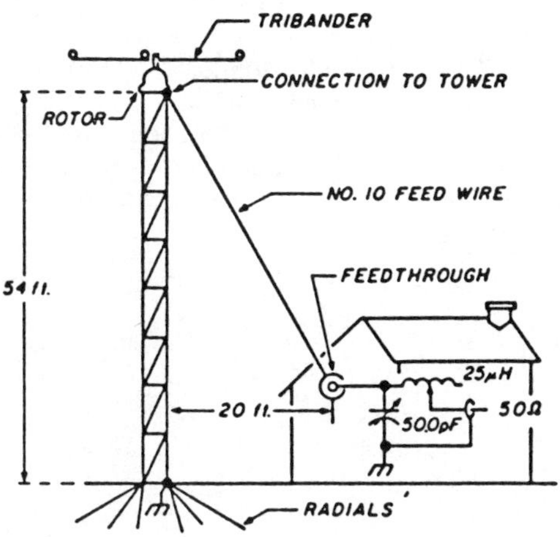

Fig. 8 160 METER FOLDED UNIPOLE makes use of 54 foot tower at W5LDA. Wire is connected to tower at top and voltage fed at the bottom. Base of the tower is grounded and twenty 65' radials fan out on ground surface. Each rotor lead is bypassed to tower at the top with a .01 μF disc capacitor. Leads are also bypassed to ground at the tower base. The L-network matches unipole to 50 ohm coax running to operating position. Antenna also operates on 80 meters by retuning network.

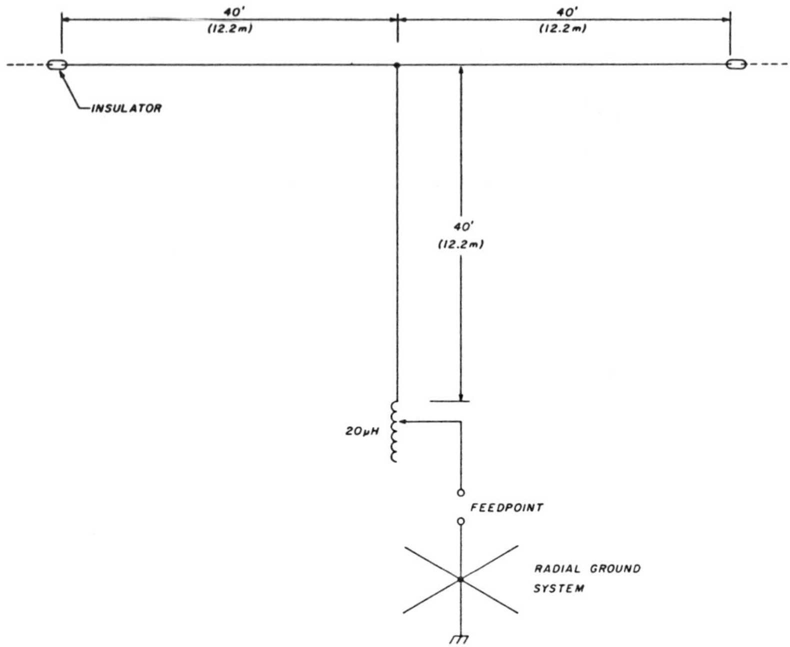

Fig. 9 COMPACT T-ANTENNA for 160 meters. 80-foot horizontal wire is center-fed and resonated to operating frequency by means of rotary inductor. Ground system consists of four radials run on surface of ground. Each radial is 40 feet long and is grounded by 3-foot rods driven into ground at 10-foot intervals along the wire. With a good ground system in average earth, feedpoint impedance is about 30 ohms.

Chapter 10

Work DX With an "Invisible" Antenna

(Unobtrusive Antennas for Apartment Dwellers and Unlucky Hams Living Where Antennas are "Prohibited")

"Modern living" has its drawbacks as well as its blessings. Some amateurs, to their sorrow, have found out that many chic townhouses and condominium apartments in the larger cities have strict rules prohibiting ham antennas. Other operators have discovered iron-clad restrictions against the erection of a permanent antenna written into the fine print of a residential property deed or a rural lease. In many instances, the amateur antenna is merely tolerated or the erection of the antenna becomes the signal for various complaints of television interference from suspicious neighbors. And other operators have found to their chagrin that space simply didn't exist in their small apartment complex to erect a full size outdoor antenna!

Antenna woes are many but usually with tact and ingenuity, an amateur may erect an unobtrusive antenna without running afoul of the neighbors, the landlord or the Building Inspector.

The legal and social problems sometimes encountered in erecting an amateur transmitting antenna (considered by a large portion of the non-amateur population to be an eye-sore) may be formidable and this Handbook does not attempt to deal with such complex questions. If a doubt exists in the reader's mind about the *legal* aspects of erecting an antenna, he should consult an attorney and get the facts *before* he erects his sky wire.

Provided there are no legal obstacles, it is possible to make the best of a delicate situation by following the principle of the "invisible" antenna. This idea works on the theory that if an antenna is not easily seen or recognized, it will not be an antagonistic object to the observer. In many hardship cases, it is possible for an amateur to erect an "invisible" antenna, get on the air and enjoy rag-chewing and

DX without anyone else being the wiser. (Needless to say this theory will not work if the transmitting gear is not fully television, telephone and stereo proof!)

This chapter provides some tried-and-true invisible antenna designs and suggestions for the amateur looking for an unobtrusive antenna. In addition, case histories of successful invisible antenna installations made under difficult situations are discussed. These may give you some clever ideas for your own invisible antenna.

The "Invisible" Antenna Concept

The invisible antenna concept is built upon the fact that the antenna is either *hidden* from view, is visible but *disguised,* or that it *disappears* from view when not in use. One of these styles of antenna can allow an amateur to get on the air under circumstances that would prohibit a more orthodox antenna installation.

There's no magic about an invisible antenna. It works according to accepted antenna theory. The only precaution is that such an antenna must be properly tuned since it may not be quite as efficient as a big outdoor antenna and every radiated watt helps! The use of an SWR meter and a radial ground wire are mandatory in most instances for proper operation. Properly adjusted, the invisible antenna can give you many hours of rewarding ham operation, regardless of the jaundiced eye of the landlord or next door neighbor!

Surveying Your Antenna Location

Before you erect any antenna, make sure that you can do it in a quiet manner without attracting undue attention. Many invisible antennas are erected at night or early dawn when inquisitive eyes are asleep or busy with other matters. Don't call attention to yourself or what you are doing and keep your cool!

The first step to take in deciding what kind of antenna is most suited to your location is to examine your residence to determine if you should use an indoor or an outdoor antenna. If the sky wire is placed indoors, it is hidden from view of the public, which may be of primary importance. Contrary to myth, an indoor antenna will do a good job, provided it is not within a building having a steel framework, walls or roof. The author, for example, has a vacation retreat in an apartment where antennas are prohibited. However, using an indoor dipole and a 200 watt transmitter, DXCC has been worked and many happy hours of operation have been logged, without the landlord or neighbors being aware that a ham station is even in the building! Even Sherlock Holmes couldn't find the antenna!

If the building has appreciable metal in the framework, the use of an indoor antenna should only be attempted as a last resort, as the structure of the building shields the antenna to a marked degree.

If it is decided to place the antenna outdoors, the antenna should either be unobtrusive, be disguised, or should disappear when not in use. For disguise, a vertical antenna can be hidden in a tree or may take the form of a flagpole. If the antenna is made of fine copper wire and very small plastic insulators and is erected high in the air, it will be nearly invisible to the casual observer. Number 28 enamel magnet wire, for example, is invisible at a distance of about 25 feet.

The vertical antenna also lends itself to the disappearing act. It can be hinged at the base so that it lies down flat against the roof when not in use, or it may be made collapsible so that it can be taken apart when not used. These ideas, and others like them, allow you to erect an antenna and enjoy ham radio in most uncompromising situations, provided you go about your ham activities without attracting undue attention.

The Indoor Antenna

The indoor antenna is height-limited by the roofline or ceiling of your dwelling. If you plan such an antenna, it is much better for the antenna to be located on an upper floor than on the ground floor of a building, if this is possible. In a wood frame building, the indoor antenna seems to perform as well as an equivalent outdoor antenna provided the indoor antenna is not accidentally coupled into the electric wiring or metal plumbing system of the building.

Survey your residence and determine its construction. The majority of homes and two story apartment buildings in the United States and Canada are of wood frame construction with a roof of composition material or wood shingles. Large multi-story apartments, on the other hand, almost always have a steel framework which usually rules out the use of an indoor antenna.

The wooden frame building seems to cause little or no effect upon the antenna provided some care is taken not to let the antenna couple itself electrically with the wiring and plumbing. If you have an attic space, a loft, or a high ceiling the antenna should be strung up near the ceiling (provided the building does not have a metal roof). Take time to determine the construction of the roof before you place an antenna beneath it!

If you have access to the roof, a dipole antenna may be laid directly on the surface of the roof or perhaps strung a few inches above it using existing vent pipes or chimneys for tie points. If #28 enamel wire and small diameter coaxial line are used, such a doublet antenna is nearly invisible, especially if the line is painted to match the color of the roof.

Some of the larger radio distributors carry RG-174/U mini-coaxial 50 ohm cable. This is only 3/32-inch in diameter and is capable of handling up to 500 watts SSB or c-w up to 30 MHz if the SWR on it is low. It is not quite invisible, but it is the nearest thing to it! When invisibility is important, RG-174/U should be used in place of the larger diameter RG-58/U or RG-8/U cable.

If it is decided to place the antenna indoors, it need not be made invisible in most instances. Depending upon the layout of your home, either a horizontal dipole, a vertical ground plane or an end-fed wire may be used on the higher frequency bands. On the lower bands, space restrictions usually limit the indoor antenna to the end-fed type. In any case, the use of a radial ground wire with the antenna to establish an electrical ground point (as discussed in Chapter 8) is *essential*. A lot of TVI and stero-QRM can be caused by the omission of this very important station accessory. If it is not used, the missing ground connection is made up of the electrical wiring of the building which includes all the TV receivers, stereo gear, radios, telephones and other electrical gadgets in the vicinity. And that's not good!

Antenna Layout

Place your indoor antenna at the high point of your dwelling and align it so that it is at right angles to most of the electric wires and pipes in the walls of the building. The position of wires and pipes, while hidden from view, may be estimated by observing the position of floor plugs, outlets and the various water faucets and drains. The antenna is installed as far away from these as possible and at right angles to them, if feasible. It is perfectly OK to let the antenna slope or to bend it a bit to fit it into the available space. The whole operation is cut-and-try and it is easy to move the antenna about to determine the optimum placement. If you are using a dipole and one side is higher than the other, the high side should be connected to the center conductor of the coaxial line and the lower side to the outer shield, as shown in Figure 1, providing no balun is used.

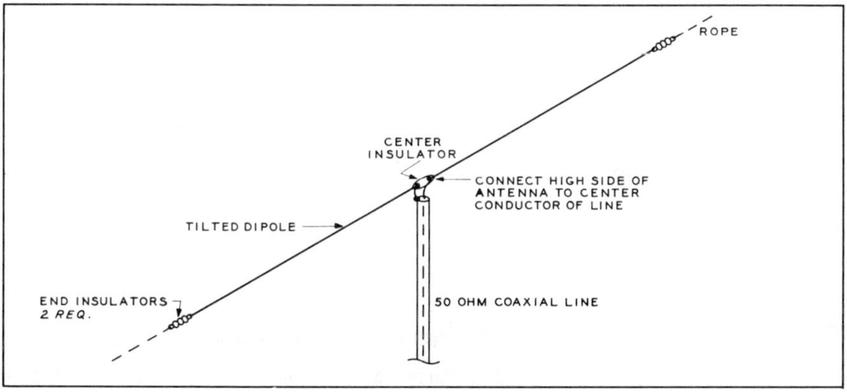

Fig. 1 SLOPING DIPOLE INSTALLATION. When one side of dipole is higher than the other the high side should be connected to the center conductor of the coaxial line and the low side to the outer shield, as shown above.

Checking the Antenna

Once the location has been chosen, the antenna is strung in place using heavy twine and hook-eyes in the building structure. The insulated radial ground wire is laid along the baseboard of the radio room, following the floor line around the room, or under a rug. The antenna is tuned and loaded in the normal manner, using an SWR meter (and tuner, if required) in the coaxial line to the antenna. It may be necessary to change the physical placement of the antenna in the room in order to arrive at a suitably low value of SWR across the band.

Once the antenna seems to be operating properly with a reasonable value of SWR, it is a good idea to check the electrical wiring of the building for r-f power that may be sneaking into it from the nearby antenna. If the wiring is encased in metal conduit, this coupling is unlikely. Open wiring (knob-and-tube or *Romex)* may act as an unwanted pickup antenna and absorb great quantities of your precious output power. Sometimes r-f coupling can be noticed when a light or lamp (supposedly turned off) lights up with a weak glow when you are on the air, or a nearby TV set chokes up and refuses to work properly. A handful of .001 uF, 1.6 kV disc capacitors can help to solve this problem. A fast and simple way to connect these capacitors is to obtain a few 115 volt line plugs and connect a capacitor across the screw terminals of each plug (Figure 2A). The plug is then inserted in wall receptacles at random until a position is found that "detunes" the electric wiring system and reduces absorbed r-f to a minimum. Sometimes two or three such plugs are required to "detune" a residence.

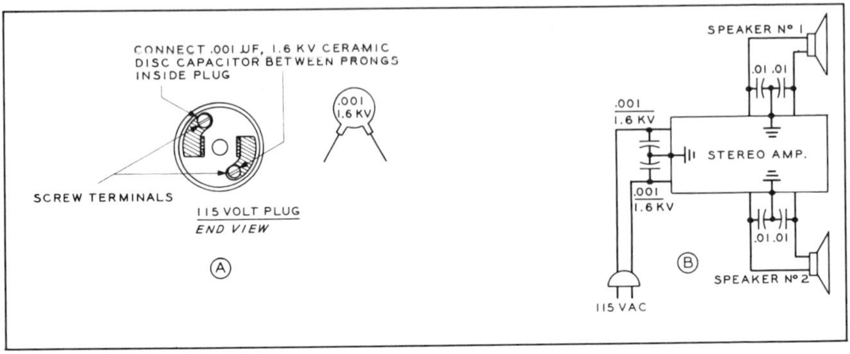

Fig. 2 **STEREO AND TV INTERFERENCE can be substantially reduced by proper r-f filtering of power lines and speaker cables. (A) Ceramic capacitor in power plug reduces r-f in power line. (B) Speaker wires and power line of stereo amplifier are bypassed to common ground return of amplifier. In many cases shielding the speaker leads also helps to reduce r-f pickup.**

In addition, it may be necessary to place an extra bypass capacitor across some of the leads of a stereo or TV that is afflicted with interference. In particular, stereo speaker leads should be bypassed to the chassis or ground return of the stereo set with .01 uF disc capacitors to prevent the wires from working as unwanted pickup antennas (Figure 2B).

What type of antenna to use indoors? Well, for operation on the 6, 10, 15 or 20 meter bands, a dipole antenna is recommended along with a radial ground wire connected to the transmitter. A tri-band dipole will work well for 20, 15 and 10 meter operation when used with three radial wires. A dipole is really too long for an indoor antenna installation on the 40 and 80 meter bands in most residences so an end-fed wire is suggested, used in conjunction with an antenna tuner, as described earlier in this Handbook.

When space is at an absolute premium, a mobile whip antenna mounted on a window ledge may be used in place of the wire antenna, or the *pole lamp* antenna described in the following section will do the job for you.

The Pole Lamp Antenna

Are you living in a studio apartment or other mini-space accomodations? A ham rig can be squeezed into a closet or bookcase but what about the antenna? Can the antenna blend in with the decor of the room and not be an eye-sore? The answer to all these questions is "yes" if a pole lamp antenna is used.

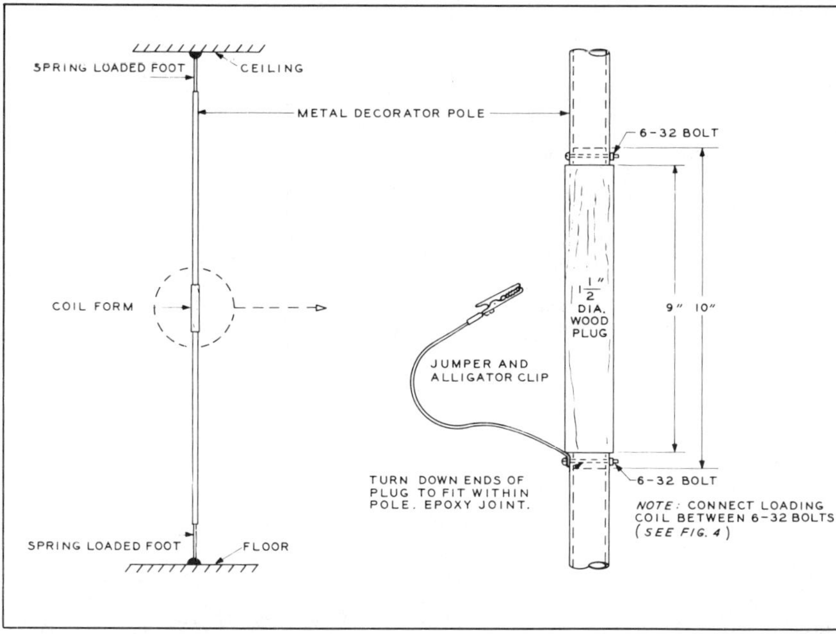

Fig. 3 POLE LAMP ANTENNA provides good multi-band operation in minimum space. Decorator pole assembly is cut in half and insulating plug inserted to provide mounting area for loading coil. Pole antenna is used in conjunction with radial ground wire for best results.

The pole antenna is a decor-matching accessory to the home if carefully built. This unique antenna is made from a spring-loaded pole lamp structure, which is placed in position between floor and ceiling of the radio room. Many home furnishings stores carry a variety of pole lamps and decorator poles which support shelves, radios, book racks and other objects. Plain, unadorned poles and the various pieces which go together to make up the poles can also be purchased. The most popular and least expensive pole is a three section assembly about seven feet long having spring-loaded "feet" in the ends which are wedged against ceiling and floor.

On occasion, a pole may be purchased with a wood center section, which is required for this antenna assembly. If only an all-metal pole is obtainable, the center section must be cut in half and trimmed for insertion of a wooden plug to serve as insulator and coil form (Fig. 3).

The spring-loaded pole antenna is operated in a vertical position, with a high-Q, tapped loading coil placed at its center. The coil tunes the compact antenna to the band in use and the complete pole antenna plus the wire connecting it to the antenna tuner are used in conjunc-

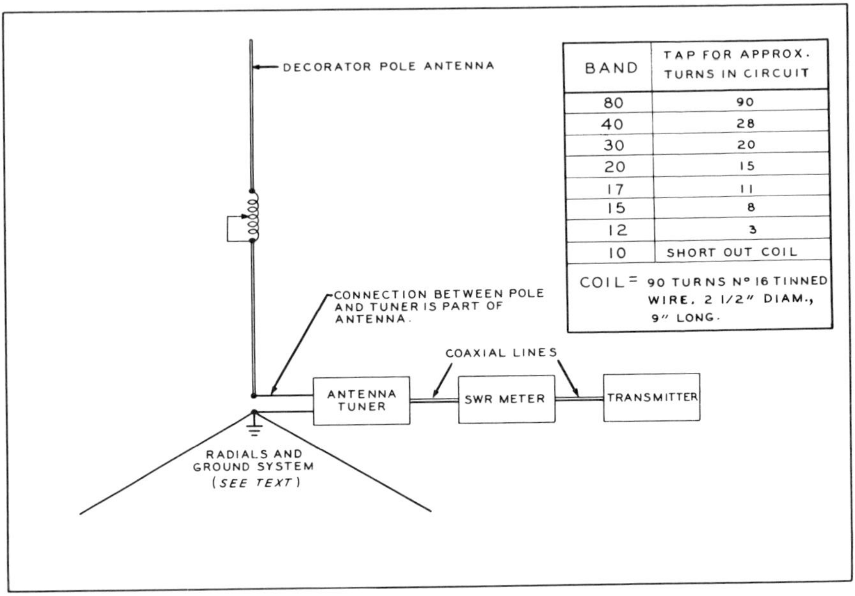

BAND	TAP FOR APPROX. TURNS IN CIRCUIT
80	90
40	28
30	20
20	15
17	11
15	8
12	3
10	SHORT OUT COIL

COIL = 90 TURNS N° 16 TINNED WIRE, 2 1/2" DIAM., 9" LONG.

Fig. 4 POLE ANTENNA operates on all h-f bands 80 through 10 meters. Tapped loading coil, antenna tuner and SWR meter allow operator to adjust the antenna for minimum SWR at the operating frequency. Quarter wavelength radial ground wire for band in use is required, as discussed in text.

tion with a radial wire, if convenient, or with a ground composed of the cold water plumbing system of the building.

A simple antenna tuner (described in Chapter 8) is located near the transmitter along with an SWR meter in the coaxial line, as shown in Figure 4. The number of turns in the loading coil of the pole antenna are adjusted so that the tuner can allow a low value of SWR on the coaxial line to the transmitter.

Some operators may raise their eyebrows at using a seven foot high, coil loaded antenna on the lower frequency bands but it should be pointed out that the efficiency of an antenna of this size is at least equal to that of the eight foot loaded mobile whip antennas that continually prove their worth on the 80 meter band.

Antenna Adjustment

The number of turns on the center loading coil is given in Figure 4 and the antenna resonated by means of the auxiliary tuner. If a reasonable value of SWR cannot be achieved, the number of turns in the center loading coil is adjusted a turn at a time until the SWR drops to a low value at the operating frequency. Any major shift in frequency (10 kHz or so) requires retuning of the antenna tuner for minimum SWR on the coaxial line to the transmitter.

"Invisible" Outdoor Antennas

If space permits, it may be feasible to string up an invisible outdoor antenna away from the building. This will provide good DX results for you. The easiest antenna to put up is an end-fed single wire which is resonated to the operating frequency by an antenna tuner located at the transmitter. The wire may be of almost any length as the tuner described in Chapter 8 will compensate for the length used. For general operation between 40 and 10 meters, the wire should be at least 30 feet long. For 80 meter operation as well, the wire should be at least 50 feet long. Lengths up to about 150 feet will work OK. In every case, an indoor radial ground wire placed at the antenna tuner should be used.

The name of the game is to make the outdoor antenna invisible, or nearly so, to the casual observer. Enameled wire of #28 gauge (magnet wire) is a good starting point. It is quite fine, yet has good tensile strength. Wire sizes as small as #34 (removed by the mile from a defunct power transformer) have been used by experimenters with good results. Operation of the antenna seems to have little to do with wire size, although some antennas made with #34 wire have mysteriously come down during the night. The only reason for this, it is guessed, is that birds can't see the wire either, and fly into it. Antennas made with #28 gauge wire do not seem to have this problem.

Very small insulators for the invisible antenna can be made out of 1/4-inch diameter lucite or plastic rod. The insulator can be about 2 inches long with fine holes drilled in it to accept the antenna and tie-off wire.

The ends of the invisible antenna should be tied off to points that will not impose a strain on the wire when the wind blows. Trees are poor tie-off points as they sway in the wind. An invisible wire run from one building to another, or to a sturdy tree is a good installation. A flag pole also serves as a good tie-point for the antenna.

In order to see the invisible antenna when you are putting it up, tie a long thread to a small piece of paper. Fold the paper and drop it over the center of the antenna wire. Once the antenna is up, you can remove the paper with a gentle tug on the thread.

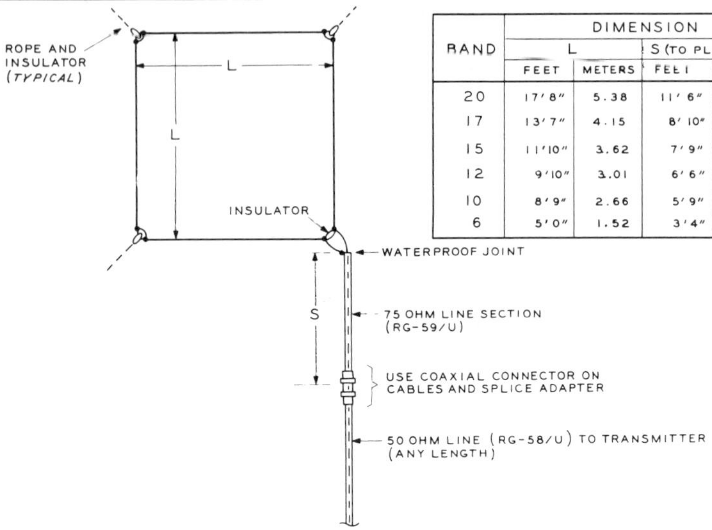

BAND	DIMENSION L		S (TO PLUG TIP)	
	FEET	METERS	FEET	METERS
20	17′ 8″	5.38	11′ 6″	3.55
17	13′ 7″	4.15	8′ 10″	2.72
15	11′10″	3.62	7′ 9″	2.39
12	9′10″	3.01	6′ 6″	2.00
10	8′ 9″	2.66	5′ 9″	1.75
6	5′ 0″	1.52	3′ 4″	1.00

Fig. 5 QUAD LOOP makes effective "invisible" antenna. Loop is made up of #28 enamel wire supported with midget insulators cut from lucite rod. The support ropes are nylon fish cord. Matching section and transmission line are brought down side of building or hidden in shrubbery or tree. Loop provides power gain of about 2 decibels over a dipole antenna.

What Antenna To Use?

A number of antenna types lend themselves to the invisibility concept. Here are some suggestions that may help you:

The Dipole. An invisible dipole can be made of #28 enameled copper wire for the flat top, using plastic rod insulators. The feedline is made of small diameter RG-174/U mini-coaxial line. The jacket of this line, unfortunately, is dark black so that it tends to stand out against the sky. It is quite unobtrusive against a darker object such as a building or a tree.

The Quad Loop. The Cubical Quad Loop element makes an effective antenna having about 2 decibels power gain over a dipole. The radiation pattern of the loop is similar to that of a dipole, being a figure-8 at right angles to the plane of the wires.

In order to remove the feedline from view, the loop is fed at either an upper or lower corner via a 75 ohm quarter-wavelength matching section and a 50 ohm transmission line, as shown in Figure 5. While the loop has low visibility, the line and transformer are more notice-

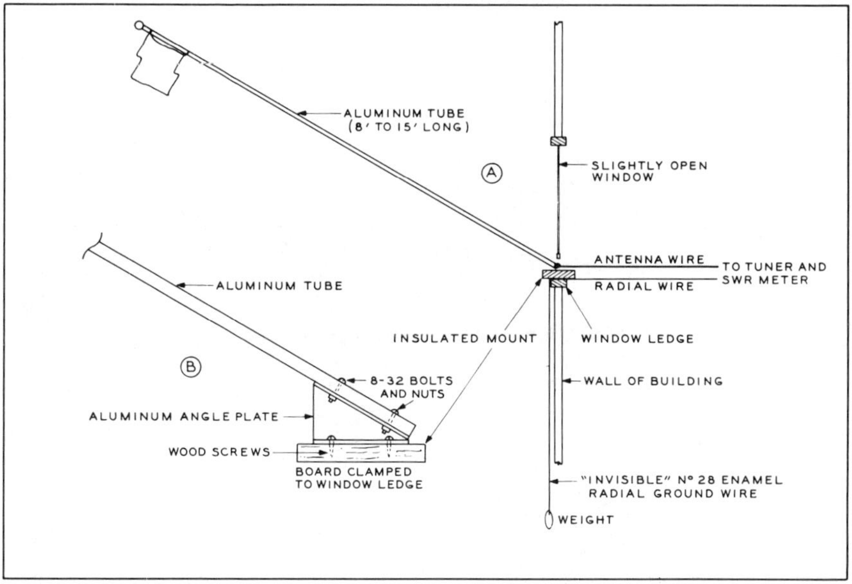

Fig. 6 PATRIOTIC FLAGPOLE is really a disguised antenna. Pole, plus wire running to antenna tuner form antenna. An "invisible" radial ground wire is dropped out of the window. The flagpole is insulated at the base if the window frame and ledge are metal. Some amateurs who have a porch or balcony with a metal railing along it use the railing for the ground system.

able and should be disguised by surrounding objects. For example, ivy, woodbine or other creeping plant may be trained to grow up the side of a building to mask the presence of a feedline.

The Dick Tracy Flag Pole Antenna

If the invisible antenna doesn't seem to be the solution to your problem, perhaps you should consider the visible, disguised antenna as an alternative design. A flag pole, for example, combines high visibility and good neighborhood acceptance. Properly built, it can serve as an excellent multi-band antenna for the high frequencies, and nobody but the radio ham is any the wiser! When viewed by the public, the antenna seems to be a simple patriotic flag pole (complete with halyard, pulleys and flag) but in fact it is a disguised transmitting antenna!

One form of flag pole antenna is shown in Figure 6. This disguised antenna is mounted to the window ledge of an apartment house or home. The pole is a length of 1-inch diameter aluminum tubing or TV mast

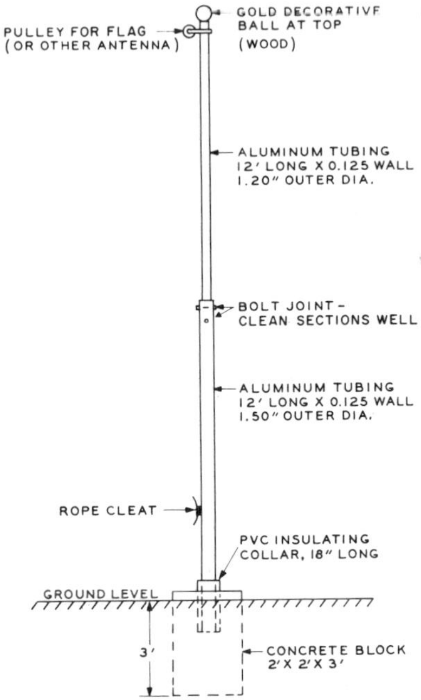

Fig. 7 FULL-SIZE FLAGPOLE serves as efficient h-f vertical antenna. The aluminum pole is insulated at the base and imbedded in a concrete block to make it self-supporting. Radial ground wires run through nearby shrubbery disguise function of the flagpole. Rope cleat at base and a gold decorative wood ball at the top complete the illusion.

and can be any length from 8 to 15 feet, depending upon aesthetic circumstances and the patriotism of the neighbors. The pole is insulated at the base from the window ledge by an insulator and is end-fed via a random length of insulated wire and the end-fed antenna tuner described in chapter 8. A single radial ground wire either inside or outside the building is used with this effective flag pole antenna.

A more complex flag pole antenna is shown in Figure 7. This is the *Dick Tracy* version, a full-fledged, patriotic flag pole, mounted on an impressive base. It would look well in the front yard of the White House! The pole is made of two sections of 1/8-inch wall aluminum tubing and can be from 15 to 20 feet high. The pole sits in a poured concrete base about two feet square and three feet deep in the ground. The flag pole radiator is insulated from ground by the concrete and also by a collar made of plastic *(polyvinyl chloride or PVC)* pipe. This inexpensive material is available at any large plumbing supply house. Buy a piece about 18" long and of a diameter that provides a loose slip fit over the lower section of the aluminum tubing. One end of the

PVC pipe is sealed with a plug of plastic or PVC sheet epoxied in place. The sealed end is sunk in the concrete to make a fully insulated socket for the antenna.

The concrete is poured into a hole dug in the ground. Make sure that the plastic socket is in a vertical position before the cement hardens!

Place the antenna in the socket and shim the space around it with thin strips of aluminum and then fill the remaining gap with dry sand, well tamped down. Finally, seal the whole joint between socket and antenna with *General Electric* RTV-102 sealant to keep moisture out.

It is a good idea to plug both the top and bottom openings of the aluminum tubes that make up the antenna. This will prevent moisture from running down the inside and filling the bottom of the antenna with water which will gradually corrode the inside of the tubes. The splice between the two sections of tubing should be waterproofed, too.

Again, the height of the pole is an aesthetic rather than an electrical consideration, and pole heights as high as 40 feet will work well for the high frequency bands. If the base is partially obscured with small bushes or a privet hedge, one or two horizontal radial ground wires can be run through the plants, about a foot above the ground. Several ground rods (obtainable from an electrical supply house) can be driven into the soil at the antenna base and connected in parallel with the radials, if convenient.

A small tuning unit such as described in Chapter 8 is placed at the base of the flag pole antenna to permit multi-band operation. It should be enclosed in a water-proof box and the connections between tuner and antenna and ground should be reasonably short. It is possible, moreover, to remotely control the tuning unit from the station position for deluxe all-band operation, but such a device is beyond the scope of this Handbook.

The coaxial feedline from the antenna and tuner can be buried underground. The exposed end of the line, of course, should be properly sealed to prevent moisture from entering the line, as discussed earlier.

The "CIA Special" -- a Disguised All-Band Antenna System

Unlucky amateurs may sometimes find to their chagrin that a ham antenna is barred from the roof of an apartment or rental home, whereas a television antenna is perfectly OK to erect! Often it is easier for the amateur to accept such a restriction than to oppose it, especially if it is possible to disguise the ham antenna to look like a TV antenna.

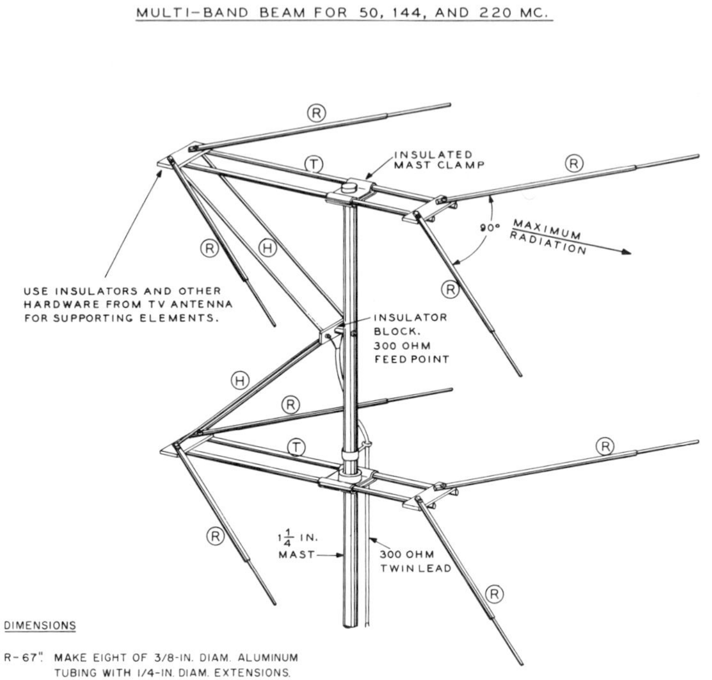

DIMENSIONS

R-67". MAKE EIGHT OF 3/8-IN. DIAM. ALUMINUM TUBING WITH 1/4-IN. DIAM. EXTENSIONS.

T-62" MAKE FOUR OF 1/2-IN. DIAM. ALUMINUM TUBING SPACED 3-1/8 IN. CENTER-TO-CENTER.

H-57" MAKE FOUR OF 1/4-IN. DIAM. ALUMINUM TUBING SPACED 4 IN. CENTER-TO-CENTER.

APPROX. FORWARD GAIN OVER HALF-WAVE DIPOLE		FRONT-TO-BACK RATIO	SIDE LOBES
50 MC.	7 DB	10 DB	NIL
144 MC.	12 DB	18 DB	-12 DB
220 MC.	13.5 DB	18 (*EST.*)	-8 DB

Fig. 8 "CIA SPECIAL" BEAM operates on 50, 144, 220 and 432 MHz bands. Array is supported on TV mast by insulated clamps near center of the horizontal phasing lines (T). Beam may be used on h-f bands by tying feed wires together and operating system as a random length wire antenna. For vhf operation, the tuner shown in Figure 9 is used.

One VHF enthusiast did this, and ended up with a VHF beam antenna and an all-band antenna for the lower frequency bands, to boot! Here's the story and a description of the antenna:

The Two-Bay VHF Beam Antenna

Shown in Figure 8 is a broadband, two bay, double-V antenna of a type often used for long distance TV fringe reception. This version is called the *CIA Special* because of its effective "cover" disguise, for it really is a ham antenna that *looks* like a TV antenna. Designed after the popular TV array, this special antenna is a highly directive beam which functions on the 50, 144, and 220 MHz bands. It may also be used as an end-fed antenna for the high frequency bands from 10 to 160 meters by treating the feedline as a radiating long wire!

Electrically, each bay of the beam is a pair of half-wave dipoles (R-R) cut to 50 MHz, one pair spaced 1/4-wavelength in front of the other. The dipoles are driven with a phase difference of 90 degrees by the horizontal phasing lines (T). This feed system produces a unidirectional radiation pattern with good front-to-back ratio and power gain of about 4 decibels over a dipole. At 144 MHz, the antenna functions on its third harmonic and begins to operate as a short V-beam, providing about 9 decibels power gain. At 220 MHz, the V-beam is working at the 5th harmonic with a power gain of about 10.5 decibels over a dipole.

Two of these beams are stacked to provide an additional 3 decibels of gain on each band and are fed in phase with a linear matching transformer (H). The two bays are separated a half-wavelength at 50 MHz, which is equivalent to 1-1/2 wavelengths at 144 MHz and 2-1/2 wavelengths at 220 MHz. Total array gain is thus the sum of the stacking gain and the bay gain, or 7 decibels at 50 MHz, 12 decibels at 144 MHz and 13.5 decibels at 220 MHz. The antenna also provides worthwhile gain over a dipole at 432 MHz, although the pattern is almost nondirectional.

Physical spacing between the upper and lower bays is determined by the positioning of the matching transformer (H) and is about 74 inches. Feedpoint impedance is close to 300 ohms on the 6 and 2 meter bands and about 200 ohms on the 220 MHz band. This impedance variation and the resultant line loss are not serious if the 300 ohm ribbon TV line is less than 75 feet long, or so. For a longer run, it is best to use open wire TV line. A suitable VHF antenna tuner for the antenna, regardless of the line used, is shown in Figure 9.

Low Frequency Operation (10 to 160 Meters)

This VHF beam antenna is used for low frequency operation down to 160 meters by ignoring the antenna and considering the feedline as a random length, end-fed antenna. For this mode of operation, the wires of the two-wire line are tied in parallel at the station end and the whole assembly considered as a single wire. An antenna tuner and a radial ground wire must be used in conjunction with an SWR meter, as discussed earlier in this Handbook.

Since the lead-in is used as the antenna on the low bands, it should be spaced away from the building and nearby metallic objects.

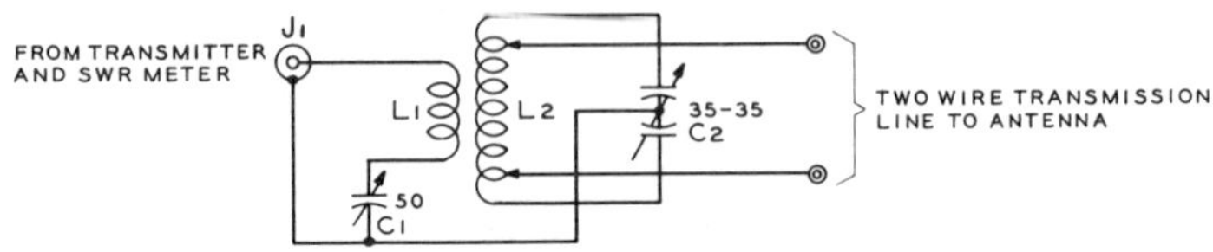

COIL DATA

50 MHz
L1- 5 TURNS N° 18, 1" DIAM., 8 TURNS PER INCH. PLACE INSIDE L2.
(*1 CORE AIR-DUX 808T OR EQUIVALENT.*)
L2- 7 TURNS N° 14, 1 1/2", DIAM., 4 TURNS PER INCH. (*1 CORE AIR-DUX 1204 OR EQUIV.*)
TAP 1 1/2 TURNS FROM EACH END.

144 MHz
L1- 2 TURNS N° 14, 1" DIAM., 8 TURNS PER INCH. PLACE OVER L2.
L2- 5 TURNS N° 12, 1/2" DIAM., 1" LONG. TAP 1 1/2" TURNS FROM EACH END.

220 MHz
L1- 1 TURN N° 14, 1" DIAM.
L2- 2 TURNS N° 12, 1/2" DIAM., 1/2" LONG. ADJUST FOR RESONANCE WITH CAPACITOR C2 NEARLY OPEN. ANTENNA CONNECTIONS AT ENDS OF COIL.

Fig. 9 "CIA SPECIAL" antenna tuner is assembled on plywood board in the manner shown in Figure 5, Chapter 12. Capacitor C_1 is Millen 19050, or equivalent. Capacitor C_2 is split-stator, 35 pF per section, 0.07" spacing; Millen 28935, or equivalent. Tuner is removed from line for operation of antenna as end-fed wire on high frequency bands.

Other Amateurs -- What Have They Done?

It is interesting to observe some real-life situations where amateurs have triumphed over local situations to get on the air. Here are a few interesting cases that may give you an idea as to how your problem can be solved:

Case History #1. The Deluxe, Flowering Flagpole.

One west coast radio amateur living in a manicured, model neighborhood placed a fine, 18 foot flag pole in his front yard, with the blessing of the City Planning Commission and the local Neighborhood Improvement Association. A flowering hedge, tenderly cared for by his XYL, hid two radial wires. The coaxial cable was run under the lawn to the house. In addition, the ground point of the antenna tuner was connected to the lawn sprinkling system.

A few months of operation proved the worth of the antenna, so our Hero inserted a three foot extension section in the flag pole at the base, bringing the height of the antenna up to 21 feet. At the same time a remotely switched antenna tuning unit was installed in the bushes at the base of the tower to provide automatic bandswitching over the 10, 15 and 20 meter bands. Shortly thereafter, a switchable base loading coil for 40 and 80 meter operation was quietly added (at night) to the box at the base of the tower.

Finally, one quiet evening, the flag pole was lowered and a two meter ground plane made of thin, silver plated steel wire was installed at the top of the pole and the coaxial line run down the inside of the pole. To cap the climax, the enterprising amateur ran #28 wire from the peak of his house roof out to the pole for use on 160 meters! Thus, in the space of a few months, he gained a multi-band antenna capable of operation from 2 meters through 160 meters, without the neighbors ever realizing that he had a ham station in the house! His amateur friends viewed the various manipulations with awe, and urged him to place an invisible 20 meter beam atop the flag pole. "That will take a little more time," he said quietly.

Case History #2. The Disappearing Ground Plane Antenna

Joe Ham lived in a housing complex that frowns upon antennas of any kind. Even the TV signal is piped in through underground conduit. The area around Joe's home was open and erecting an invisible long wire was impractical. The home was a single story affair and an indoor antenna would not be very high in the air. Obviously, the situation called for drastic action.

After due deliberation, Joe decided upon a 20 meter ground plane antenna using four radial wires lying on the flat roof of the house. The vertical aluminum tubing radiator of the ground plane was hinged at the base and designed to lie flat against the roof when not in use. Joe reasoned correctly that he was at work all day, spent most of the weekend with his family and friends, and the only time he was on the air was during the very early morning hours or late at night. At these times he reasoned that the antenna could be erected with little risk of discovery.

Accordingly, Joe laid the radial wires on the roof and attached the base of the vertical antenna to a hinged board. At first, Joe mounted a ladder to get to the roof level. He pulled on ropes to raise and lower the antenna. Later, a remote control TV rotator was purchased, mounted on its side atop the roof and used to erect and lower the vertical radiator. Joe has had the antenna in operation for almost two years now and no word has been said, no voice raised in opposition, and no problems with the neighbors. "No sweat," said Joe, when interviewed by the author. "I'm on the air, working DX, and everything is calm in Shady Acres."

Case History #3. The Collapsible Antenna

Pendergast lived in a single family dwelling in an area which boasted of cable TV, underground utility wires and no radio antennas of any kind. The mere thought of a sky wire was *verboten* to the neighbors. Pendergast, being an ardent traffic handler and 40 meter enthusiast, designed a take-apart vertical antenna made up of 8 foot sections of 1-1/4" diameter telescoping TV mast. Four sections made a nifty 32 foot high vertical antenna that required no guy wires when anchored to the side of his house. The supports for the antenna were bolted to the side of the house away from the street and just outside the radio shack window.

Pendergast did most of his operating rather late in the evening, and threw in a few hours of DXing on 80 meters before sunrise. Accordingly, a forty and an eighty meter pair of radial ground wires were laid out in the yard about a foot above the ground, passing at random through flower beds and small hedges thoughtfully planted at strategic points by Pendegast's XYL. The 8 foot mast sections could be quickly assembled into a vertical antenna that fitted into the house brackets in a jiffy. The antenna acted as a ground plane on 40 meters and a simple loading coil at the base of the antenna permitted 80 meter operation. One of these days Pendergast is going to build an end-fed tuner and operate his antenna on 20 meters. Meanwhile, he enjoys 40 and 80 meter SSB operation with an occasional fling at c-w. "I have WAC and over 50 countries on 40 meters and 3 continents on 80 meters so far this season," says Pendergast. "Not bad for an antenna that isn't there most of the time!"

Case History #4. The Antenna in the Tree

Tommy Trueheart's father, a crusty old gentleman, was fiercely proud of his prize-winning rose garden which took up almost all of the space in the backyard of the modest home. He forbade any foolish ideas about wireless antennas disrupting the placid calm of the beautiful rose garden. "I'll have none of that tom-foolery," he said, stroking his moustache as he surveyed his roses.

Tommy, ever persevering, approached his father when the old geezer was in an amiable mood, admiring the trophies and badges he had won in garden competition. After a long discussion of the beauties of the garden and attributes of various rose species, Tommy turned the conversation to ham radio and antennas. Finally, grudging permission was granted to permit an unobtrusive antenna to be placed in a large, leafy oak tree at the side of the home, *provided* the father did not set his eye on the wires. A visible antenna, the old codger warned, was equivalent to having the ham license revoked by the FCC!

Being a dutiful son, Trueheart obeyed and soon had the old tree strung with a tri-band dipole and a Quad loop, both made of insulated hook-up wire and strung between handy branches. Quite a bit of DX was worked on 15 and 20 meters with this rat's nest of wires. In fact, Trueheart's father came into the shack on occasion when the son thoughtfully contacted an amateur interested in rose gardens.

Trueheart was apprehensive when winter came and the leaves gradually fell from the tree and the antenna wires were exposed to the naked eye. However, time had played its game, and the old curmudgeon no longer looked up into the tree each day to see if he could spot the incriminating wires. In fact, after a long chat with an amateur in Lima, Peru, about roses, the old boy suggested to Trueheart that a modest tower and antenna in the corner of the yard would obviously produce stronger signals from Peru than a bunch of wires in an oak tree! Needless to say, Trueheart took the hint and now a tri-band beam on a slim crank-up tower reposes in the far corner of the pretty rose garden.

Case History #5. No Antenna At All!

On the air with no antenna at all? One amateur did just that! No outdoor antenna was permitted, and the XYL objected violently to any indoor antenna. It seemed as if Roger Grommethead was off the air for good. A close investigation of the house, however, showed that a galvanized metal gutter ran around two edges of the roof and was connected to two down-spouts near the ends of the house. In desperation, Grommethead ran a small wire out the window and soldered it to a downspout. The ham station fitted unobtrusively in a corner of the hall closet. Using an antenna tuner, the down spout and gutter pipe system was successfully loaded up for 80 and 40 meter operation, using a rod driven into the soil outside the window for a ground. Operation on 20 and 15 meters, however, brought near-disaster. TVI was rampant and the signals overrode the stereo and blocked out the telephone. Grommethead placed a filter on his transmitter and also on the TV, bypassed the speaker leads to the stereo, hounded the telephone company to place a filter on the phone *and* added 20 and 15 meter radial ground wires to the antenna system. He ran the insulated wires around the baseboard of the room. Interference was cleaned up and now this pleased amateur works four ham bands, any time of day, with no antenna at all (to speak of)!

OK -- It's Up To You!

This chapter has given you some ideas about antennas to use when it is difficult or impossible to erect a conventional sky wire. Circumstances are different in almost every case, and the antenna that suits one amateur cannot always satisfy another. Here is where ingenuity pays off! Remember that any metallic structure can serve as an antenna of some sort -- including bed springs! If the structure takes power from the transmitter and does not overheat while doing so, it is a pretty good indication that a large portion of the power is being radiated into space.

Your invisible antenna is limited only by your imagination. One amateur built a 6 meter beam out of cardboard tubing and aluminum kitchen foil, which was wrapped about the tubing. Another ham made a 10 meter dipole out of strips of aluminum foil stuck to two large windows with rubber cement. Wire dipoles and simple beams may be placed in unused attic space. Rain gutters can act as an antenna, and a mobile whip antenna can be projected out the window at an angle, in conjunction with a radial ground wire.

So you *can* get on the air in any location -- with a little bit of imagination, ingenuity and courage! Good DX to you!

Howy Bradley, W2QHH, Hamilton, NY, is known world-wide for his DX exploits. He has worked 364 countries (105 on 160 meters) with 20 watts output to an end-fed single wire 264 ft. long and 40 ft. high. Skillful operating, patience, persistence, and an antenna that works paid off for W2QHH.

DECIBEL CHART

Decibel (dB)	Power Gain
1.0	1.26
1.5	1.29
2.0	1.58
2.5	1.78
3.0	1.99
3.5	2.24
4.0	2.51
4.5	2.82
5.0	3.16
5.5	3.55
6.0	3.98
6.5	4.47
7.0	5.01
7.5	5.62
8.0	6.31
8.5	7.08
9.0	7.94
10.0	10.00
20.0	100.00

In antenna work, power gain (or loss) is expressed in decibels. Thus, a Yagi beam with a gain of 7.5 dB will have a power gain of 5.62 compared to a dipole (which has a power gain of zero). This means a 100 watt transmitter will have the "punch" of a 562 watt transmitter. An antenna gain of 10 dB will make a 100 watt transmitter equivalent to a 1000 watt transmitter using a dipole.

Chapter 11

Wire Beam Antennas for DX!

(You Can't Work 'em if You Can't Hear 'em)

Plenty of easy DX can be heard and worked with a dipole or ground plane antenna and many amateurs have worked DXCC using these simple and inexpensive antennas. The serious amateur soon will wish to use a beam antenna, however, to improve his signal in competition and to help him hear the rare "weak ones" which may be inaudible unless he is using a "gain" antenna.

Shown in this chapter are simple wire beam antennas that are fun to build and are inexpensive and uncomplicated. They are great projects for the amateur who has gained experience erecting wire antennas and scrambling over rooftops!

Build a Demi-Quad Beam Antenna for 20, 15 or 10 Meters

The *Demi-Quad* antenna is a compact and inexpensive beam antenna that can be supported by a single pole or mast. It is light, unobtrusive, and has the same "figure-8" bidirectional radiation pattern through the loop as exhibited by a dipole antenna. The Demi-Quad may be installed so as to have either horizontal or vertical polarization by proper placement of the feedline, as discussed later. The antenna provides about 2 decibels power gain over a dipole and requires only a half-turn for complete coverage of the compass. It is light enough to be supported and turned by a TV rotator.

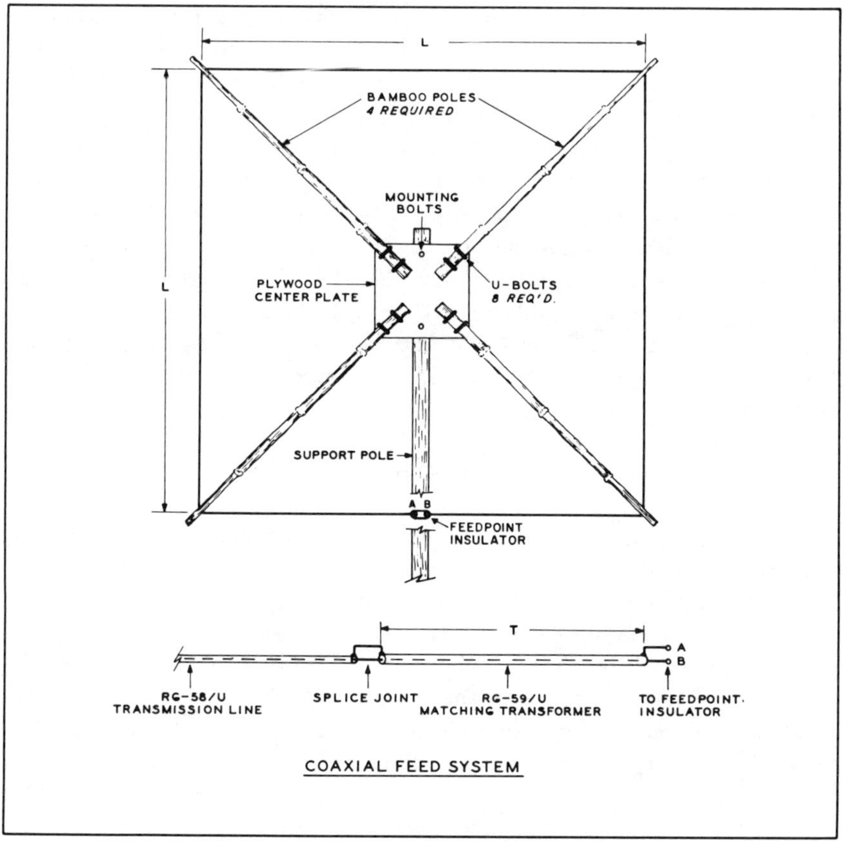

Fig. 1 DEMI-QUAD loop antenna is inexpensive and simple bidirectional beam that is supported from one pole. Compact and light, this popular antenna is recommended for 20, 15 or 10 meter operation. Demi-Quad uses simple coaxial matching transformer to provide low SWR on feedline.

Demi-Quad-Assembly

The complete Demi-Quad loop antenna is shown in Figure 1. A light bamboo frame supports the wire loop in the vertical plane. Each side of the loop is about one quarter-wavelength long, and the loop is broken at one point for the 50 ohm (RG-58A/U) coaxial transmission line and a simple matching transformer made of 70 ohm coaxial line. If the line and transformer are attached to the loop at the middle of the bottom section, the antenna is horizontally polarized. If the connection is made at the middle of one side (either side), the antenna is vertically polarized.

The framework of the antenna is assembled from four lengths of bamboo attached at their large ends to a plywood center plate by means of galvanized iron U-bolts. Each pole is wrapped with vinyl plastic tape between joints to enhance the strength of the assembly and to retard splitting of the bamboo. In addition, each pole is given several coats of waterproof varnish after wrapping. Small holes are drilled as shown near the tips of each pole to pass the antenna wire which is threaded through the poles after assembly of the framework. Each end of the wire is cleaned and the ends are passed through the holes of the center insulator, wrapped back upon themselves and soldered. Enough tension may be imparted to the wires to keep them taut by loosening the center U-bolts and spreading the butt ends of the poles.

The Feedline

The 50 ohm coaxial transmission line must be long enough to reach from the transmitter to the center plate of the antenna. At the antenna end of the line, connection is made to the antenna terminals through a short section of 70 ohm coaxial line (RG-59/U) which acts as a matching transformer. This transformer matches the 50 ohm line more closely to the radiation resistance of the Demi-Quad antenna, which is about 120 ohms. It is only necessary to cut the 70 ohm line to the proper length for it to make this impedance transformation (Figure 2).

Antenna Construction

The first job is to cut the center plate out of 1/2-inch thick plywood and give it several coats of outside house paint. Pay special attention to the edges of the plywood, as water will attack the glue if the edges are not well painted. Drill the plate for the U-bolts and temporarily assemble the bamboo poles to the plate. Mark the exact center of the plate and measure out the distance to the wire on each arm *from the center point*. Mark the distance on each pole. This is the spot where you will drill a hole in each arm through which to pass the antenna wire. Stretch the wire out into a straight line, and temporarily attach it at the marks by means of a piece of string and some tape. It should be a tight fit. If it seems too loose, you may have to mark a new hole a *little* farther out on one or two poles than the position you have just marked. When you have found the correct points, drill each pole carefully with a drill just slightly larger than the wire size. Now, pass the wire through the holes, attach the center insulator at the

	L (SIDE LENGTH)		S (DRILLING POINT)		T (TRANSFORMER LENGTH)	
	FEET	METERS	FEET	METERS	FEET	METERS
20	17′ 8″	5.38	12′ 6″	3.81	11′ 8″	3.57
17	13′ 7″	4.15	9′ 8″	2.96	9′ 0″	2.75
15	11′ 10″	3.62	8′ 4″	2.53	7′ 11″	2.41
12	9′ 10″	3.01	7′ 1″	2.16	6′ 8″	2.04
10	8′ 9″	2.66	6′ 3″	1.90	5′ 11″	1.80

1 — DIMENSIONS MEASURED FROM DRILL HOLE TO CENTER OF SUPPORT PLATE

2 — DIMENSION T INCLUDES 1″ LEADS AT EACH END OF TRANSFORMER

Fig. 2 DIMENSIONS OF DEMI-QUAD ANTENNA

bottom of the antenna and tighten things up by pushing the butt ends of the poles out a bit in the U-bolts. Don't change the wire length; that determines antenna resonance.

Antenna Polarization

Antenna polarization depends on how you mount the Demi-Quad to the support pole. For vertical polarization, the center insulator should be at the side of the antenna. For horizontal polarization, the insulator should be at the bottom of the antenna. You can experiment with polarization, if you wish, by swinging the antenna from a horizontal to a vertical position, achieving both vertical and horizontal polarization at the same time when the bamboo poles form a vertical cross and all wires are tilted at a 45 degree angle.

The Coaxial Feed System

As mentioned before, the feedline consists of a 50 ohm transmission line plus a special short section of 70 ohm line that acts as a form of matching transformer. The 70 ohm line should be cut to a length which allows 1½ inches at each end to make connections. The line ends are prepared with pig-tails, as shown in Figure 2, Chapter 7. Take care in making the splice as it is easy to melt the polyethylene center insulation of the line with the heat of the soldering gun. To make the splice, twist the center conductors together and solder them carefully. When cool, wrap the joint with vinyl tape, continuing the tape over the polyethylene insulation at each end of the splice. The braid pig-tails are now carefully twisted together and soldered. The last step is to wrap the completed joint with two layers of vinyl tape, overlapping the windings as you go to make the wrapping waterproof. A liberal coating of waterproofing *General Electric RTV-102* compound completes the joint.

You are now ready to attach the free end of the 70 ohm coaxial matching transformer to the antenna at the center insulator. The line should be supported from the center pole so that the weight of the line does not pull at the antenna joint. Tape the line to the pole and center plate. The center conductor of the line is soldered to one end of the antenna loop and the outer conductor pig-tail to the other end of the loop. The pig-tail should take the strain if the line is pulled. When connections are completed, the end of the line is wrapped and waterproofed with *RTV-102* compound.

Antenna Installation

The Demi-Quad antenna is quite light and may be handled by one small ham even though the wires may have a devilish tendency to tangle with nearby objects! If a metal support pipe or mast is used, take care that the coaxial line and bottom of the antenna do not bang against the metal.

Up she goes! The antenna may be turned by hand (the "Armstrong" method) or by an inexpensive TV rotator. Since the pattern of the antenna is bidirectional and very broad, it is really only necessary to turn the antenna about 180 degrees to obtain complete coverage. Two ropes attached to the lower crossarms will easily turn the antenna and also hold it in position so that it will not be turned by the wind.

The Demi-Quad operates across an entire ham band. The SWR may be measured with the usual 50 ohm SWR meter and readings will run between 1.3 and 1.7 across a band. The flat surface of the loop is aimed in the direction you wish to receive or transmit.

A Quad Beam Antenna for 6 Meters (Great for FM or SSB!)

A two element Quad beam antenna provides a power gain of nearly 8 decibels over the conventional ground plane antenna. This is certainly a worthwhile power boost and is useful in both receiving and transmitting. Weak signals will pop up out of the background when a beam of this type is used as compared against a ground plane and the transmitted signal will "run rings" around the ground plane, too.

Described in this section is a simple design for a two element, vertically polarized Quad beam antenna for 6 meter SSB or FM work. It will really bang your signal into a distant repeater or permit simplex

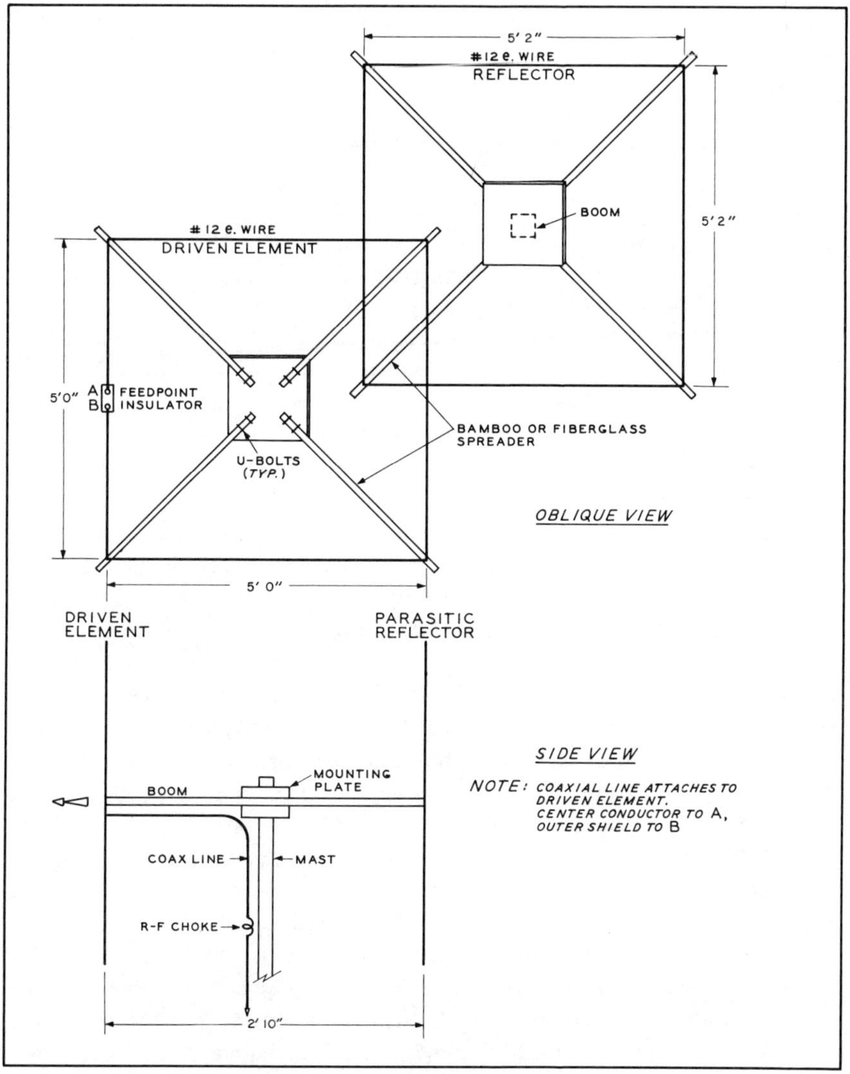

Fig. 3 QUAD BEAM antenna provides 8 decibels power gain over 6 meter ground plane antenna. Completed beam resembles the design shown in the photograph on page 22. Quad loops are made of wire strung around a bamboo frame. Driven element is fed at the side for vertical polarization or at bottom for horizontal polarization. Quad is fed with 50 ohm coaxial transmission line at feedpoints A-B, with center conductor of line connected to point A and outer shield connected to point B. The line is coiled into an r-f choke, as shown in Figure 4.

operation over a long range. The SSB DXer, moreover, can use the beam for improved sporadic-E DX and long distance ground wave coverage.

Beam Antenna Assembly

A simple and rugged Quad assembly is constructed of bamboo support arms and a wood center plate, as described in the previous section. The two element Quad uses two separate loops, one acting as the driven element and the other as a parasitic reflector, as shown in Figure 3. The directivity of the beam is *from* the reflector *through* the driven element. The loops are supported by means of a short horizontal boom which may be made of either wood or metal. Each loop is assembled in the manner described for the Demi-Quad antenna.

The two element Quad is a true beam antenna and has a unidirectional (one way) pattern, showing a good front-to-back ratio. It is therefore necessary to rotate the beam through 360 degrees in order to obtain complete radio coverage. A heavy duty TV rotator will do the job.

Antenna Assembly

The Quad loops are assembled in the manner described for the Demi-Quad. Note that the reflector loop is somewhat larger in size than the director loop and the mounting holes for the antenna wires are drilled a bit farther out on the bamboo poles. It is suggested that extra-length poles be used so that the small tips may be cut off and discarded. Bamboo poles, by the way, may often be purchased at bamboo distributors in large cities, at some rug stores in smaller towns and sometimes at garden nurseries and hardware stores.

The boom should be made of a section of dry 2'' x 2'' lumber, well painted to protect it from moisture in the air. ''Green'' lumber tends to warp as it dries out, imparting an unlovely twist to the symmetrical Quad antenna. Use dry lumber, sand it well, and give it two coats of outdoor house paint.

The center plates of the Quad loops are attached to the ends of the wood boom by means of four galvanized steel angle brackets. The brackets are mounted slightly off-center on the boom so that the retaining bolts will not interfere with each other passing through the boom. Do not use wood screws at these joints as they probably will work loose in the first wind storm. The completed wood and bamboo structure seems to have about as much structural strength as a jellyfish. However, once the wires are strung in position and made taut, the

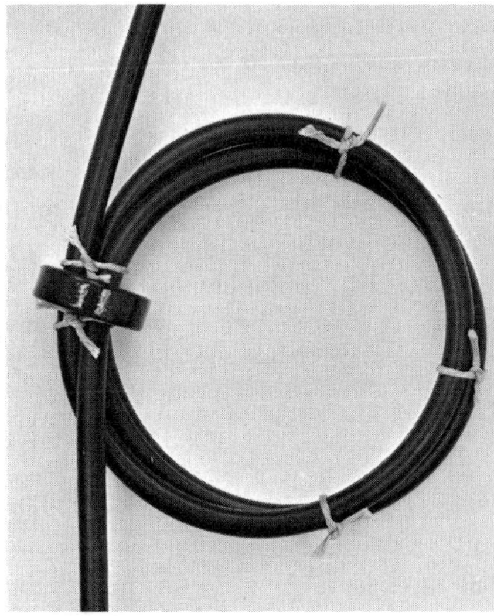

Fig. 4 FERRITE R-F choke coil is made by winding coaxial transmission line around core. RG-58/U line is wound into 4 inch diameter, 2 turn coil. Core is Indiana General CF-117. For data, catalog and list of distributors, write: Indiana General Corp., Crow Mills Road, Keasby, N.J. 08832. Core is Q-1 material, 1.875" diameter.

assembly magically becomes strong and amazingly rigid. Believe it or not!

Note that the reflector element has no center insulator and is just a complete circle of wire. The driven loop, as in the Demi-Quad, is broken by an insulator at the feed point. Positioning this loop will determine whether the beam is horizontally or vertically polarized, as described in the previous section.

The final assembly operation is to bolt the loop assemblies to the boom. A little pre-planning at this point is helpful, because once the Quad is assembled it becomes an unwieldy object. A good idea is to place the boom atop a six-foot step ladder to keep the whole antenna in the clear above ground during the final assembly stages. Do not lift the Quad by the bamboo arms as this tends to warp the assembly.

Once the beam is completed, the 50 ohm transmission line should be attached. The line is connected directly to the antenna ends at the center insulator of the driven loop. No matching transformer is required. Previous remarks about waterproofing the end of the line should be remembered at this stage of the game.

The Ferrite R-F Choke Coil

The 2 element Quad has a good front-to-back ratio and good gain. In order to make sure that the presence of the feedline does not upset the electrical characteristics of the Quad, an r-f choke coil is placed in the coaxial feedline. This device prevents r-f energy at the antenna from passing down the *outer surface* of the coaxial line. Remember: all energy must remain *inside* the line. To make a suitable choke coil, the coaxial line is merely passed twice through the center hole of a small ferrite core, making two loops of the line about 4 inches in diameter as shown in Figure 4. The choke coil is positioned about 8 feet down the line from the antenna. After looping the line through the core, tape the coils to the core so that the turns will not move about. It is a good idea to fill the inside of the core with cloth and then tape the whole assembly as the ferrite material is quite fragile, and may shatter if it bangs against the mast on windy days.

Antenna Installation

Normally, you'll build the Quad antenna for vertical polarization with the feedline brought away at one side of the driven element loop. The line should be dressed inwards towards the boom, then along the boom to the center support, and then dropped down the mast to the station. Tape the line to the boom to keep it from flopping around in the wind. If you use a TV rotator, you'll have to leave enough slack in the line so that the rotor will not twist the line into a knot when the antenna turns or be stopped by too short a length of cable. Mount the Quad so that the lower wires of the loop are at least ten feet above any structure, such as a roof. As with any antenna, best results will be obtained with the Quad as high in the air as legally possible, and clear of other objects. The SWR on the transmission line will run between 1.2 and 1.6 across the 6 meter amateur band when the antenna is in the clear.

A "LONG JOHN" Yagi Beam for 2 Meters

Dollar for dollar and pound for pound, the "Long John" Yagi beam has consistently outperformed other types of beam antennas for both local and long distance work on the 2 meter band. Huge antenna arrays made up of many individual Yagi beams are used for moonbounce experiments and long distance DX by some of the experts in the VHF world.

Here's the design for a high gain, 2 meter beam made of aluminum clothesline wire and wood you can build that will out-perform other beam antennas costing more than twice as much to buy. Best of all, this design requires no adjustment and is sure-fire in performance.

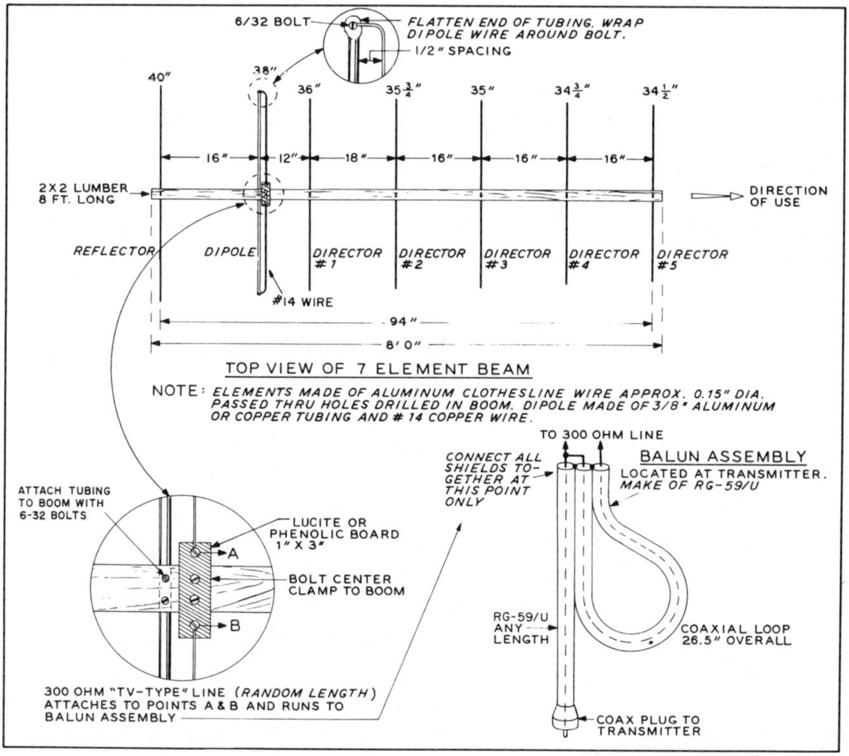

Fig. 5 SEVEN ELEMENT YAGI BEAM for 2 meter band provides 11 decibel power gain over dipole. Beam may be mounted either horizontally or vertically depending upon polarization desired. Dimensions shown are for 144 to 146 MHz. For operation over 146 to 148 MHz, all element lengths should be reduced by one-half inch.

The Antenna Design

A seven element Yagi beam providing about 11 decibels power gain over a dipole can be built on an eight foot boom. Shown in Figure 5 is an antenna of this type which uses a folded dipole, a reflector and five directors. Element lengths and spacings are chosen for maximum power gain consistent with good coverage of the 2 meter band. Contrary to most designs, this antenna is fed with a 300 ohm "TV-type" ribbon line and employs a simple half-wave balun transformer *at the station* to provide a proper match to most VHF equipment which is designed to be used with coaxial transmission lines.

The reason the ribbon line is used is that coaxial line has appreciable r-f loss at 2 meters, one hundred feet of coaxial line having nearly as much power loss as the antenna produces in power gain! Changing antenna design so that ribbon transmission line can be used drops line loss to less than 2 decibels per hundred feed of line, a figure much more reasonable and acceptable! In addition, the ribbon line is inexpensive and easily obtained at your nearby TV service and repair shop, or many hardware stores. The whole antenna, in fact, can be built of material readily obtainable at a large hardware store or home service center.

Antenna Construction

The antenna support is an eight foot boom made of 2'' x 2'' dry lumber (which actually measures about 1½'' x 1½'' in cross-section). Sand down the lumber so that it is smooth and free of splinters. Give the boom two coats of outdoor paint or varnish and drill the element holes as shown in the illustration. The holes all lie along the center line of the boom and should be drilled on a drill press if possible to insure that all elements are parallel and at right angles to the long axis of the boom. The holes should provide a tight fit for the aluminum wire elements.

The aluminum wire is now stretched tightly between two anchor points to remove kinks and bends from it. Cut it carefully into element sections of the proper lengths and carefully round the tips of the elements with a file. When they are completed, pass the elements through the holes in the boom, centering them properly so that the antenna is square and true to the eye.

The folded dipole element is made up of a 38'' length of 3/8-inch diameter aluminum tubing. The ends of the element are flattened in a vise and drilled to pass 6-32 plated machine screws. The wire portion of the folded dipole is made up of short lengths of #14 copper wire which are connected between the element tips and the insulating center block. The wire is spaced 1/2-inch away from the tubing. The center block and aluminum tube are bolted to the wood boom at the proper point and a 300 ohm ribbon line attached to the terminals of the block. The beam is now complete and ready for use.

A word of caution at this point: Don't paint the elements! A coat of paint will detune them and reduce antenna gain to a suprising degree. Leave the elements clean and bright and give them a thin coat of *Krylon* spray from an aerosol can to keep them from corroding as time passes.

Positioning the Antenna

Height is the key to successful VHF antenna operation! Vertical polarization is often used on 2 meters for FM operation in many areas of the country, whereas horizontal polarization is occasionally used, both for FM and SSB operation in other areas. Check and see which polarization is in general use in your area.

The antenna should be supported at the center of gravity by a wood mast that is long enough to remove the antenna from the metal structure of the tower, TV rotator and nearby objects. For vertical placement, the ribbon line should be run along the boom towards the reflector element and taken off from the rear of the antenna *behind the reflector,* otherwise the feedline may tend to distort the good directional pattern of the antenna. If the antenna is mounted horizontally, the ribbon line may drop down directly below the antenna.

You can mount the antenna on the wood mast described later in this Handbook, or an inexpensive TV-style "slip-up" mast may be used in conjunction with a TV-rotator.

The Balun Transformer

The conversion from ribbon line to coaxial line is done at the station, with the aid of the *balun transformer* shown in the illustration. "Balun" stands for *balance-to-unbalance,* and indicates that a balanced, two-wire ribbon line is attached to an unbalanced, coaxial line with the aid of a balancing device. In this case, the particular balun used also makes an impedance transformation from 300 ohms down to about 75 ohms, or so. This provides a good match for RG-59/U coaxial line running to the 2 meter equipment.

The balun assembly consists simply of a 26.5" length of RG-59/U line looped back upon itself and connected to the line to the 2 meter equipment at the point it joins the ribbon line to the antenna. The line to the equipment may be any length, up to 20 feet or so, before line loss becomes appreciable. It is good practice to place the balun near the transmitter and hold coaxial line length between balun and equipment to a few feet. The balun is connected as shown in the illustration, all outer shields being connected together, and one end of the balun inner conductor connected to the common junction of the coaxial line and one side of the ribbon line. The opposite end of the balun inner conductor connects to the opposite wire of the ribbon line.

SWR Measurements

This antenna is fed with a 70 ohm coaxial line running from the balun transformer to the station equipment. An SWR meter can be placed in the line to check the operation of the antenna. A special 70 ohm SWR meter is required to make meaningful measurements as the common 50 ohm variety is not calibrated for this type of line. At least one imported SWR meter is capable of working on both 50 and 70 ohm lines and suitable designs for VHF style SWR meters for either value of line impedance are described in various amateur radio handbooks.

The dimensions given in Figure 5 are for operation over the 144 to 146 MHz portion of the two meter band. For operation over the 146 to 148 MHz portion,all element lengths should be reduced by one-half inch.

"A Whole New World......"

A whole new world opens up to your station and your ears when you go to a beam antenna. In addition, your signal reports improve sharply and you find the DX easier to hear and work!

A few years ago a friend of the author finally switched over from a ground plane antenna to a modest two element Quad. After a few weeks, he reported, "I never would have believed the difference between the two antennas! I've heard and worked more DX in the past few weeks than in the previous six months with the old antenna. Instead of getting code practice calling stations that never came back, I'm now having *fun* working them!"

Well, perhaps that is an overstatement. But the majority of successful DXers use a beam antenna of one kind or another for the highly competitive DX aspect of amateur radio.

The author of this Handbook has written two other Handbooks which cover DX beam antennas from A to Z. They are:

All About Cubical Quad Antennas. The Handbook covering Quad theory, design, construction and operation. This popular book contains gain figures for Quads, correct dimensions in feet and inches for building Quads, matching and tuning procedures made easy and full data on the amazing "Monster Quad", the "king" of DX antennas.

Beam Antenna Handbook. Theory, construction, design and adjustment of parasitic (Yagi) beam antennas, including data on popular tri-band beams, compact beams for 20 and 40 meters, how to evaluate your beam, test instruments and how to use them, plus complete data on matching systems. *See back of Handbook for how to order these books.*

Chapter 12

A Universal H-F Antenna System

(Cover All Frequencies From 3.5 to 29.7 MHz With One Antenna!)

The dipole antenna discussed in Chapter 4 is a half-wavelength element fed at the center with a coaxial transmission line. This is a *great* single band antenna but it is not suited to multi-band operation.

By modifying the feed system the dipole element can be made into a truly universal antenna capable of operating on *any* frequency between 3.5 and 29.7 MHz. To achieve this wideband characteristic, the dipole antenna must be fed with a balanced, open wire transmission line operating from an antenna tuner, as shown in figure 1.

Those experimenters who want a universal antenna that provides good results on all h-f amateur bands and can be used for reception at any frequency outside the ham bands would do well to consider this interesting, inexpensive antenna installation.

The Center-Fed Antenna

The *center-fed antenna* has been used in one form or another for many years. It can be thought of as a long wire antenna having the center portion folded back upon itself to form a two wire feedline. Because each half of the flat-top portion is the same length, the current flowing in the feedline will be balanced, regardless of the operating frequency of the antenna.

Referring back to Chapter 4, Figure 3, the pattern of voltage and current waves shown in that example also apply to this center-fed antenna, much in the manner shown in Figure 2 of this Chapter. Thus,

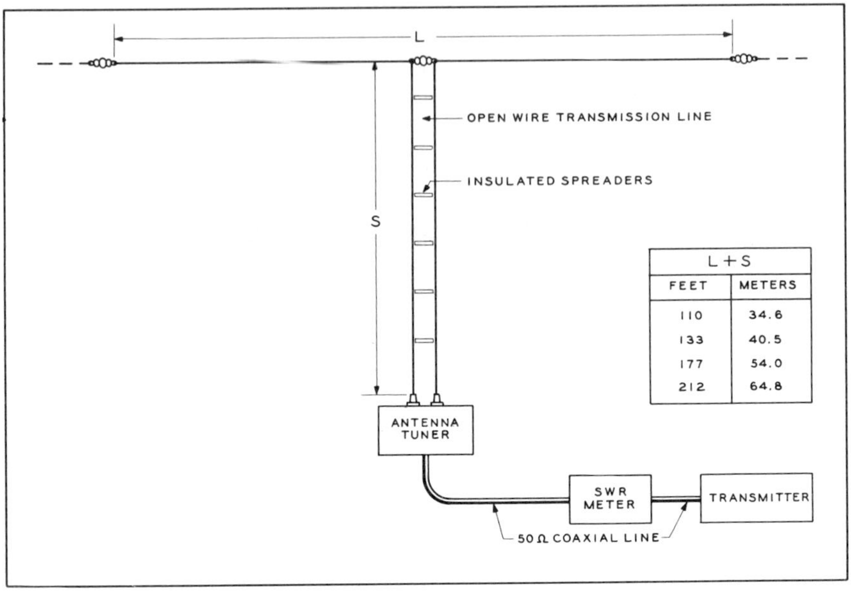

L+S	
FEET	METERS
110	34.6
133	40.5
177	54.0
212	64.8

Fig. 1 UNIVERSAL H-F ANTENNA covers 3.5 to 29.7 MHz. This center-fed antenna utilizes a simple antenna tuner (Figure 3) and open wire feedline to cover most of high frequency spectrum. Antenna and feeder lengths are not critical, and suggested lengths are given for ease in tuning adjustments. Length (L) is tip-to-tip measurement of flat-top and length (S) is length of open wire transmission line. Total length of wire in the line is twice the length (S). Length of (L) plus (S) should equal one of the four dimension combinations listed above.

the voltage wave is maximum at the ends of the antenna wire and the crest of the voltage wave falls at the low point of the current wave. In addition, these two waves are out of phase on the folded, feedline portion of the antenna so very little radio energy escapes from the feedline. The flat-top antenna does all the work!

It can be imagined that if the frequency of the radio wave applied to this antenna is varied from 3.5 to 29.7 MHz, the resulting voltage and current waves will dance about on the antenna and feedline and the resulting ratio of voltage to current measured at the bottom end of the feedline will change from a very low to a very high figure as the frequency changes--and that's exactly what takes place! In order to use the antenna at *any* frequency under these conditions, it is necessary to use a flexible antenna tuner so that the voltage to current ratio at the bottom end of the feedline can be matched to the ratio demanded by the 50 ohm transmission line running to the radio equipment.

When the tuner is properly adjusted with the aid of an SWR meter in

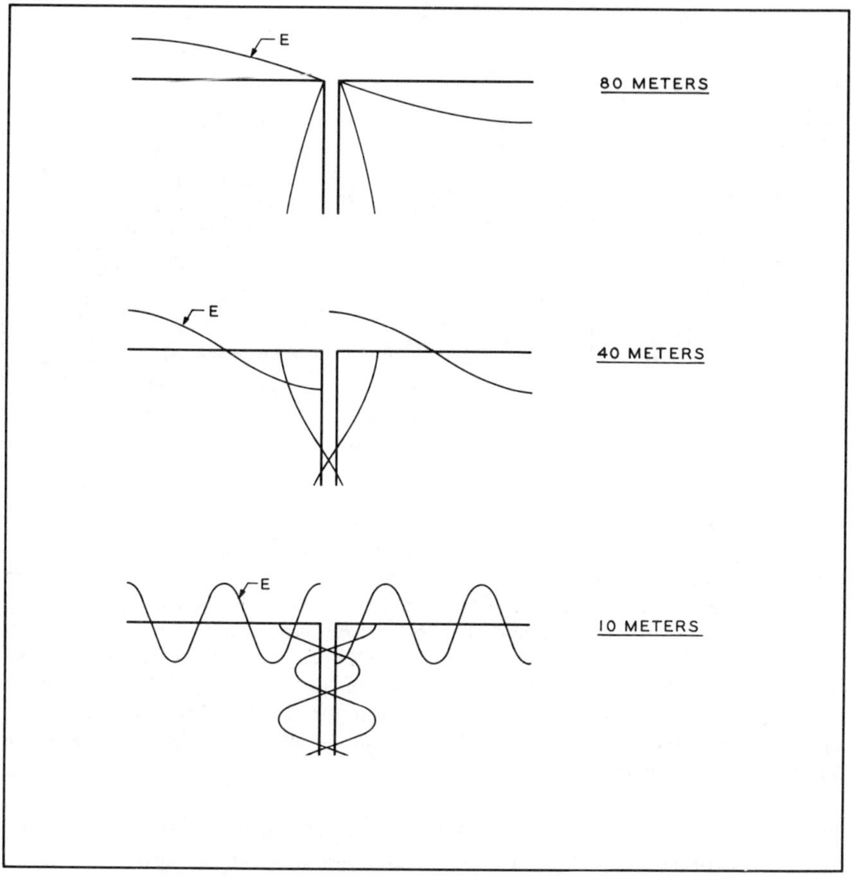

Fig. 2 VOLTAGE WAVE on universal antenna dances about as the frequency of operation is changed. As frequency increases, wave becomes more complex. Above example is for flat-top length (L) of 133 feet and feeder length (S) of 40.5 feet. Other dimension sets are given in Figure 1.

the coaxial line, a good impedance match between the center-fed antenna and the coaxial line will be achieved, regardless of the operating frequency of the station equipment.

A Practical Antenna Design

The overall length of the center-fed antenna and the length of the feedline are not critical. Some particular lengths will provide a more reasonable voltage to current ratio at the antenna tuner than will others and when these lengths are used, a less-complicated tuner can

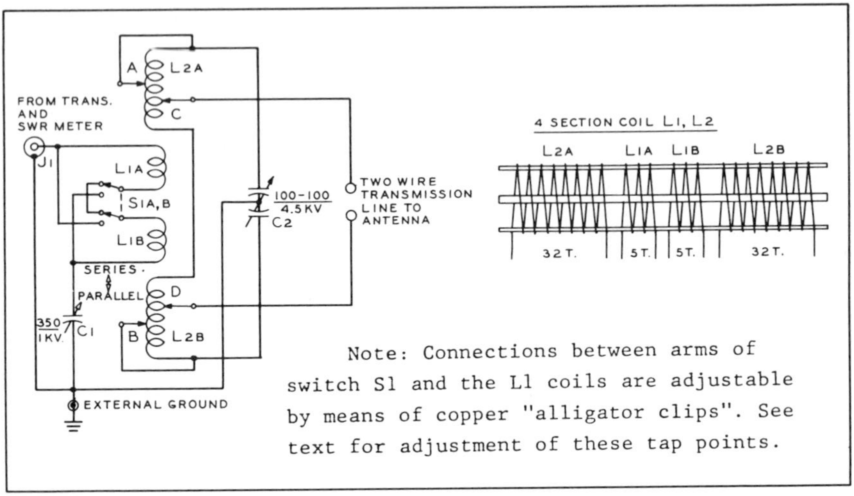

Fig. 3 UNIVERSAL ANTENNA TUNER. The four section coil is made from a single length of coil stock (Icore Air-Dux 2008 or equivalent). The coil is 2½" inside diameter, 8 turns per inch of #14 wire. Leave a 6" lead on one end and count 32 turns. Break the 33d turn at the center to make leads for L_{2A} and L_{1A}. Five more turns are counted and the coil broken at the 6th turn to make the opposite lead for coil L_{1A} and lead for coil L_{1B}. Five more turns are counted and the 6th turn broken to make the leads for coils L_{1B} and L_{2B}. Adjacent leads from the center coils are connected to arms of the switch. Coil clips are Meuller #88. Capacitor C_1 is Johnson 154-2 or equivalent. Capacitor C_2 is Johnson 154-510 or equivalent.

be built to do the job. Four suggested antenna and feeder length cominations are listed in Figure 1 that have been used by many amateurs and have proven to do a good job from 80 to 10 meters. Other lengths, determined by experiment, also may work. In addition, the total wire length may be juggled back and forth between the antenna and the feedline, with little effect upon antenna efficiency.

As an example, the first set of dimensions listed in Figure 1 call for an overall length of *flat-top plus feeder wires* of 110 feet. In general, the more antenna wire in the flat-top and the less in the feeder, the better the results will be. A flat-top length of, say, 80 feet may be selected to fit the available space, leaving a feeder length of 110 - 80 = 30 feet. On the other hand, if the user is cramped for space and the station is located some distance from the antenna site, an overall flat-top length of 60 feet may be selected, leaving a feeder length of 110 - 60 = 50 feet. If, as another example, the station is located rather close to the flat-top, a flat-top length of 100 feet may be chosen, leaving only a short feeder length of 10 feet.

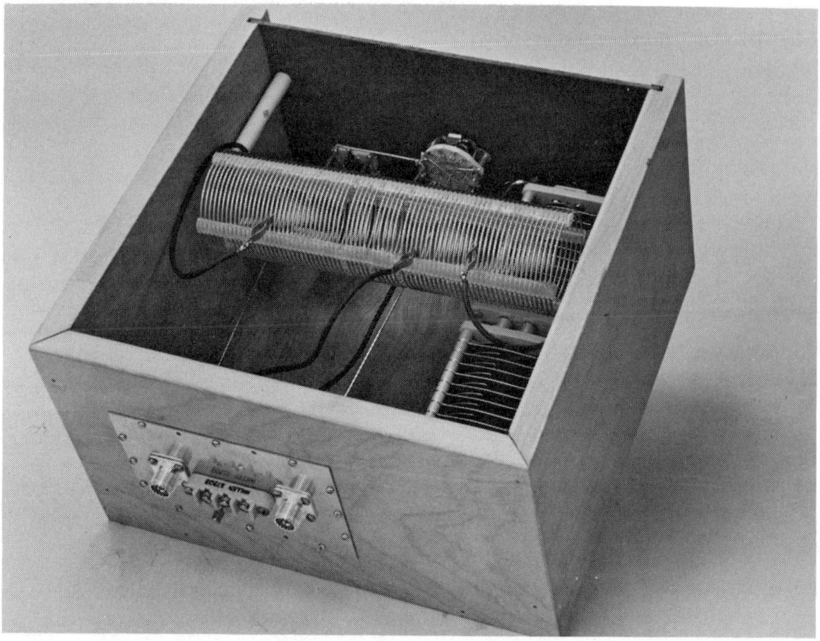

Fig. 6 INTERIOR VIEW OF CENTER-FED ANTENNA TUNER. Tapped coil and tuning capacitor permit this flexible tuner to be used over very wide frequency range. Coil turns are indented along one side so that copper clips make good contact. Coaxial input receptacle (and spare) are mounted on metal plate at rear of tuner. Antenna connections are made to ceramic terminal strip at center. Box is made of 5/8-inch wood with Masonite panel. Tuner works with almost any length flat-top antenna and feeder combination on 80-40-20-15 and 10 meter amateur bands.

You will find that various tap settings and tuning adjustments give a good degree of loading on each band and that the setting of the taps and tuning are not critical. With proper settings the transmitter loads smoothly, and the exact settings of the tuner should be logged for future use. Maximum flexibility and ability to change frequency within a band will probably be found when the tuner achieves final adjustment with both tuning capacitors about half-meshed.

"Odd-Ball" Antenna Lengths

This tuner works with almost any length flat-top and feeder combination, but the experimenter may find some random combination of wire may refuse to load on a particular frequency. In such a case, a condition of match may be established by changing the length of the feeders a few feet. It is simpler to add length to the feeders than to subtract it, and an auxiliary feeder section three or four feet long will probably cure the problem.

Caution--High Voltage

Those amateurs who have never used two wire, open transmission line with an antenna of this type may be suprised at the very high r-f voltage that appears along the line. Even with low power, r-f potentials in excess of 1000 volts may be on the line at the high potential points. Because of this, the line should not be touched during operation of the transmitting equipment. In addition, the line should not be allowed to touch nearby objects, especially metal ones. An r-f arc between the line and a metallic object severely detunes the matching network and may damage the transmitter. Properly suspended in the open air where it cannot be touched, this transmission line is very efficient and will cause no troubles.

Antenna Directivity

On the 80 and 40 meter bands this antenna exhibits little directivity and the radiation pattern is quite broad. If the antenna is very high and in the clear, some increase in signal reports may be received broadside to the antenna, as opposed to those reports received from stations off the ends of the antenna. On 20, 15 and 10 meters the antenna exhibits some directivity, with maximum signal reports noticed at an angle to the wire with a *slight* reduction in reports broadside to the wire and off the ends. This is normal, as the radiation pattern on these bands somewhat resembles a four-leaf clover. The leaves of the clover represent the radiation pattern and the space between the leaves the nulls of the pattern.

The theoretical directional pattern of the antenna, however, is usually obscured by reflection of the radio signal from nearby houses, utility wires and hills and the antenna can be considered to be virtually nondirectional except off the very ends. Thus, in the United States, if the antenna runs in a north-south direction, the user can be sure that he is covering the radio compass and that his signals will reach the eager ears of the expectant DX stations!

TV Ribbon Line

Figure 1, page 158, describes a universal hf antenna with wide frequency coverage which includes the three new amateur bands. With transmitter outputs of 100 watts or less, 300 ohm TV ribbon line may be substituted for the open wire line, if desired. To make this substitution, dimension S must be modified. First, select the overall length L plus S; then, multiply S by 0.82 to compensate for the velocity factor of the ribbon feedline. *Amphenol* or *Belden* tubular or flat feedline are recommended.

For example, assume L plus S of 177 feet (54 m) is chosen and S is to be 40 feet (12.2 m) if open wire line is used. The new length for S will be 40 X 0.82 =32.8', or 32'9" (10 m. See Chapter 13 for data on changing fractions of a foot to inches and centimeters). Final antenna dimensions, then, are L =137' (41.8 m) and S =32'9" (10 m).

It should be noted that antenna tuning adjustments might change somewhat during damp, snowy, or rainy weather when the ribbon line has a coating of water or ice.

Tuning Up on the New Bands

The Universal Antenna System (pages 157-166) will work on the three new amateur bands with no modifications other than the tap points on the antenna tuner coils may have to be readjusted. As a starter, primary coil L1 is set for 6 turns for 30 meters, 4 turns for 17 meters and 3 turns for 12 meters. Switch S1 is set for parallel coil connection.

SWR and the Solid-State Transmitter

Most modern solid-state transmitters and transceivers protect the output stage transistors by sensing the SWR of the antenna system. If the SWR is high, the transmitter reduces its output power; the greater the mismatch, the more the power is reduced.

The solid-state transmitter usually shows only a slight reduction in power output at SWR levels less than 2-to-1. Above this value, output power drops rapidly. An SWR meter is essential in monitoring the operation and determining if power output is being restricted by high SWR.

A matching unit placed between the transmitter and the coaxial line to the antenna can reduce the SWR figure to a very low value so that little or no power is lost. Such a device is shown in pages 109 and 131 of the "Radio Amateur Antenna Handbook", published by Radio Publications Inc. (See rear of book for ordering information).

Chapter 13

Antenna Round-Up

(All About Baluns, Lightning Protection, Masts -- and Lots More!)

This chapter covers a number of important antenna subjects that are generally ignored in many amateur Handbooks. However, they are of prime importance to the station operator. Of great concern to all amateurs is the problem of lightning protection, which is discussed first.

Lightning Protection

Over 400 people are killed each year by lightning according to the Census Bureau. Don't let your amateur station make you a statistic in this toll! Lightning is the most lethal of electric discharges and is entirely unpredictable.

Lightning is a tremendous electrical spark between clouds, or between clouds and the earth, releasing millions of volts at tremendous currents in a fraction of a second. The U.S. Weather Bureau indicates that Florida and the Gulf Coast have the greatest number of lightning storms per year, followed by the central states. The West coast has the smallest number of lightning storms per year.

Regardless of where you live, don't invite lightning to strike your antenna installation. A few simple precautions may save your equipment, and even your life!

In rural areas, barns and houses have been protected from lightning for many years by "lightning rods" which seem to provide a "static

Fig. 1 LIGHTNING ARRESTOR FOR COAXIAL LINE is important antenna component for all amateur stations. This small unit has precision, built-in spark gap that drains accumulated static charge on antenna to ground.

electricity drain" so that electricity in the air may pass safely to ground. This makes the immediate area around the rod less conductive, thereby diverting a potential lightning stroke to an area of greater conductivity. To accomplish the same result, your tower and ham antenna should be grounded directly, or fitted with some sort of an air-gap lightning arrestor. This device has a very small gap built into it between the center conductor and the shield of the coaxial line and, when electrical charges build up in the atmosphere around the antenna, small sparks jump the gap and drain the accumulated charge to ground (Figure 1). This type of arrestor is especially important for use with a ground plane antenna whose active element projects high in the air above the ground plane elements and is thus exposed to a direct hit from lightning.

Grounding Your Coaxial Line

While it may not be necessary from an r-f point of view it is important as a matter of safety to ground the coaxial line running from your antenna at the station end. Do this by means of a heavy strap attached to the outer, braided jacket of the line and running to one or more ground rods driven six or more feet into the soil. Rugged, aluminum grounding wire may be used, as discussed in the next section. In addition, when a lightning storm approaches the station, it is wise to disconnect the transmission line from your equipment and to remove the line from the vicinity of your station.

The ground connection insures that zero voltage exists between equipment and ground under normal operating conditions. This is important as a safety measure as it limits the voltage on the cabinet, microphone and coaxial line to a safe value, thus reducing inadvertent shock hazard. The ground connection, finally, provides an easy path to ground for static electricity and for electric currents resulting from certain equipment faults and short circuits. And, as mentioned before, the ground connection provides a static electricity drain.

If you are unlucky enough to sustain a direct lightning hit on your antenna no devices or arrestors will save the installation. The antenna and feedline will be heavily damaged or vaporized and the equipment ruined. You don't want to be in the vicinity when that happens!

Play safe! Use a lightning arrestor in your coaxial line to bleed small static charges to ground. Use a good external ground on your coaxial line. Finally, when a lightning storm approaches, disconnect the line from your equipment and, if possible, remove it from your house. Have respect for lightning and it may have respect for you!

The *National Electric Code* (NFPA #70) adopted by the *National Protection Association* (section 810-26) recommends that the grounding conductor for transmitting and receiving equipment should be connected to a metallic underground water piping system (the cold water pipe) or a metal ground rod driven at least six feet into the earth. The ground wire should be #4 gauge copper or aluminum fastened securely to the equipment at one end and the ground rod or piping system at the other. The ground lead should be as short as possible.

It should not be assumed that your metal tower or mast is grounded if it is resting on the ground or on a concrete pad. A long ground rod should be driven into the soil and the tower or mast connected to the rod with a heavy ground wire, or strap.

Build A Ferrite Balun Transformer

A *balancing transformer* (balun) is a transformer for converting a balanced electrical system to an unbalanced system, or vice versa. Balun transformers come in all sizes from the midget "ladder transformer" used with television antennas to giant multi-kilowatt units used in broadcasting stations.

Baluns are useful in antenna systems where it is desired to feed a balanced antenna with an unbalanced (coaxial) line. A balanced antenna is a symmetrical one that has two feed points, neither one of which is grounded. Dipole and nondirectional antennas, even if balanced,

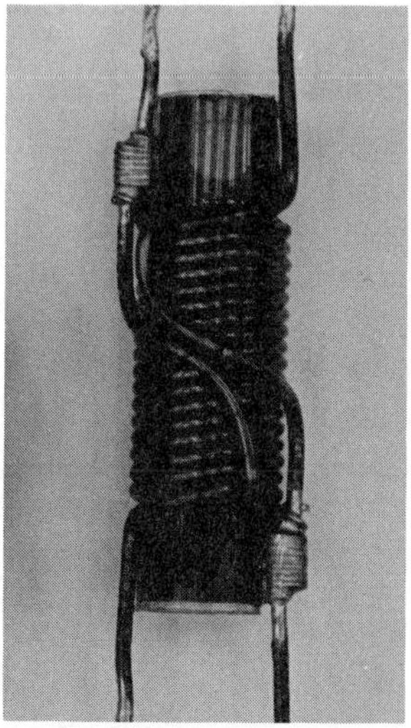

Fig. 2 FERRITE CORE BALUN TRANSFORMER is used to convert a balanced antenna system for use with a coaxial line. This inexpensive balun consists of three windings of #14 enamel copper wire wound on a ferrite slug. See Figure 3 for winding details. Balun should be mounted at the antenna terminals in a waterproof container.

generally do not require the use of a balun as its use usually does not make any great improvement in antenna operation. In the case of high gain beam antennas, however, the use of a balun is recommended. If the balun is not used, the SWR on the transmission line may rise, the SWR reading may be inaccurate, or the polar pattern or front-to-back ratio of the beam may be seriously affected. Feedline interaction of this sort is termed *antenna effect* and is most noticeable on arrays having high gain and good front-to-back ratio.

Placing a balun transformer between a coaxial feedline and a balanced antenna (at the antenna) permits the coaxial line to perform its proper duty, that of transporting radio energy from one place to another and prevents it from becoming part of the antenna itself.

A balun, or course, may be used with the various dipole antenna designs shown in this Handbook, but its use is not mandatory.

Building a Practical Balun

A simple balun transformer you can build is shown in Figure 2. It will match a balanced antenna to a 50 or 70 ohm coaxial line without disrupting the SWR on the transmission line. Power handling capability of the balun is about 600 watts average power on the 80, 40 and 20 meter bands; 400 watts on the 15 meter band and about 250 watts on the 10 meter band. The balun consists of three windings placed side by side on a short, ferrite core. This type of winding is called a *trifilar* winding.

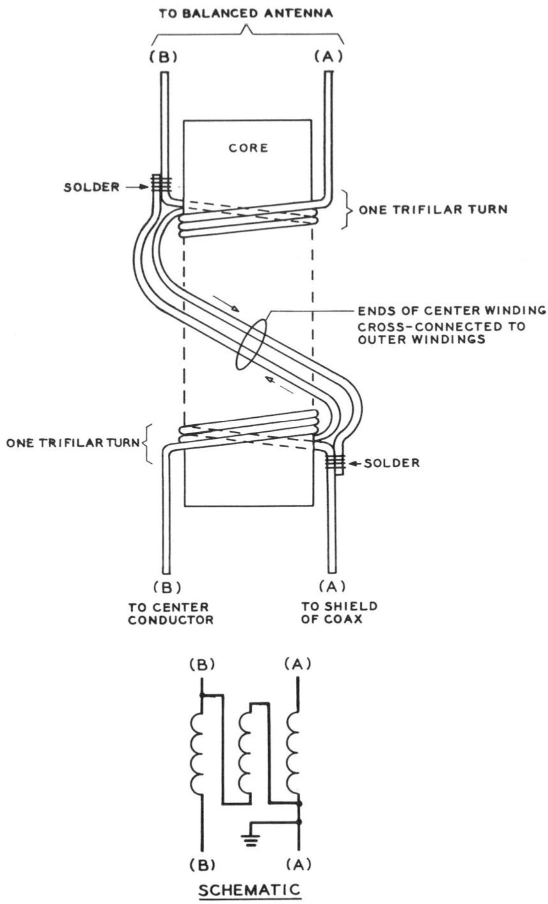

Fig. 3 WINDINGS FOR FERRITE BALUN. Three parallel windings are placed on the slug. Windings are wound on as one wire. Center winding is cross-connected to the opposite ends of the outer windings, with common connection to outer braid of coaxial line.

The balun shown in the photograph consists of three coils, each 6 turns of #14 enamel wire wound on a Q-1 ferrite slug. The ferrite slug is ½-inch in diameter and about 3¼ inches long. It is a section of an *Indiana General CF-503* slug which is 7½ inches long. The ferrite is broken to length by nicking it with a file around the circumference at the desired length and breaking it with a sharp blow. Here is how you make the special windings:

The simplest way to do the job is to make the three windings as one. Cut three pieces of #14 enamel copper wire, each about 3 feet long. Place one end of the wires in a vise and smooth out the sections until they are parallel. Grasp the free ends in your hand and wind them side-by-side on the ferrite core as if they were one wire. Wind the wires under tension and they will remain in position. When you are finished, dress the leads away from the core as shown in the photograph.

You'll note that the center winding is cross-connected to the opposite ends of the two outer windings (Figure 3). Either end of the balun may be taken as input or output, but the common connection between the inner and outer winding at the end you choose as the input end must be taken as a ground point and attached to the outer shield of the coaxial line.

Cut the ends of the inner winding to length, scrape off all the enamel and make the solder connections to the outer windings, as shown in the photograph. A drop or two of coil dope, Krylon or nail polish may be put on the ends of the windings to anchor them to the core. Do not coat the windings themselves with any material as this tends to upset the balance of the transformer.

You should protect the balun from the ravages of the weather. This can be done by placing it within a plastic bottle and coating the entrance holes with *General Electric RTV-102* or other waterproofing compound.

Using the Balun

The balun is placed at the center point of the antenna between the coaxial line and the antenna connections. Note that the balun is symmetrical end-for-end; that is, either end may connect to the antenna or to the coaxial line. The important point to remember is that the shield of the coaxial line must connect to the junction of the center and inner balun windings and the inner conductor connects to the "free" winding. This is shown in the drawing but some amateurs ignore this important point. If the connections are accidentally reversed, the balancing action is lost.

Measuring R-F Loss in Coaxial Line

The efficiency of your coaxial line may be determined by measuring the line loss at your operating frequency. You can do this by merely short circuiting the far end of your coaxial line and measuring the

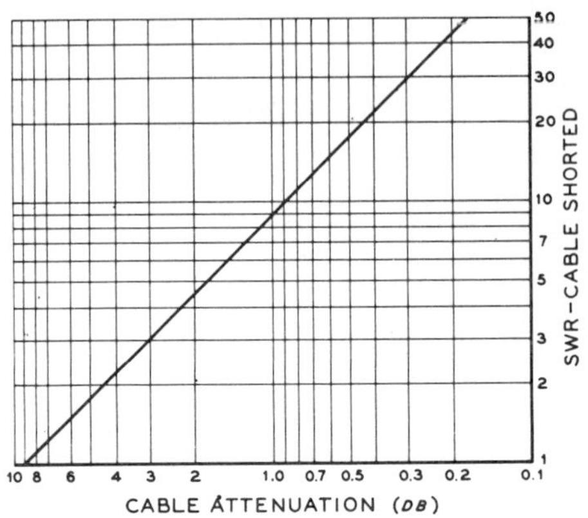

Fig. 4 COAXIAL LINE LOSS may be determined by simple test. End of line is shorted and SWR is measured under shorted condition. For example, if SWR is 10, line loss is 0.9 decibel.

standing wave ratio with your SWR meter. If there is no line loss whatsoever, the SWR reading will be infinite (full scale), indicating that the reflected wave is equal in amplitude (size) to the incident wave. In a real-life situation, of course, this is not the case, and the SWR reading under the test condition will be less than infinite, due to line loss.

In order to make this measurement, the antenna termination is removed from the far end of the transmission line and the outer shield is firmly shorted to the inner conductor of the line. A small amount of power is applied to the line through the SWR meter. The meter is adjusted for full scale reading on the "forward" position, and the meter switch is then thrown to the "reverse" position. The line loss may then be computed from the reverse reading and the chart in Figure 4. If, for example, the SWR turns out to be 4.5, the cable loss (attenuation) is 2 decibels. This means that your coaxial line is about 63 percent efficient, and that 37 percent of your transmitter output power is being lost in the line. If the SWR reading, on the other hand, is 9; then your line loss is only 1 decibel and your line is about 80 percent efficient.

The Soldering Gun

For general work on coaxial lines and connectors, a dual heat (145-210 watt), medium size soldering gun is recommended. Don't use a low wattage "pencil iron" for antenna work, as this tool doesn't have the wattage to properly heat the coaxial plugs and fittings which have a great deal of thermal mass. A large, high wattage soldering gun, on the other hand, can overheat coaxial cable and melt the inner insulation. A gun in the 140 to 250 watt range seems to be best for all-around antenna work.

If the soldering gun is new, read the instructions and prepare the tip for use. If the gun is old and the tip is dirty or has had a lot of use, prepare it by filing it with a fine file until the working surface is shiny. Then heat the gun and coat the tip with rosin core solder. *Never* use acid core solder, soldering paste or flux as the chemicals can quickly ruin the gun tip and the equipment you are working on.

The Joint

When the gun is ready to be used, the joint should be prepared for soldering. The two metals to be joined should be completely clean. Remove all dirt, enamel, scale or oxidation by sanding or scraping down to the clean metal. Insulation may be removed from wires with the blade of a pocket knife. (Make sure you do not nick the wire).

The first step is to tin the metal surfaces. Hold the hot tip of the gun against the metal until the solder melts and flows onto the clean material. When properly tinned, the metal should be covered with a thin coat of solder, and the excess solder will flow onto the tip of the gun. Apply the solder to the joint, don't try to bring it to the joint on the tip of the gun.

When you are working with two wires, or wire leads, you should make a good mechanical joint between the parts being soldered. Antenna joints should be well wrapped before soldering to overcome the stresses and strains in the wire under the continuous action of the wind.

The final step is to apply the soldering gun to the connection. Hold the tinned surface of the tip against the joint and let it come up to temperature. When it reaches the proper level, solder applied to the joint will flow quickly and smoothly over the joint. Never try to solder by applying solder to the iron and then to the joint, or by applying a lump of solder to the joint and then pressing it down with the gun.

A little experience will teach you just how hot the joint should be. If it is not hot enough, the solder will have a grainy appearance and the joint may snap apart. This is called a *cold solder joint*. If the joint temperature is excessive, you may damage the components you are working with, or melt the inner insulation of the coaxial cable. Most beginners err on the side of cold joints, as they are impatient and do not take the time to let the joint come up to the proper temperature.

Soldering Outdoors

Soldering connections on an antenna installation in cold weather can be an exasperating experience as the cold air robs the soldering gun of heat at a very fast rate. To overcome this, the easiest way to bring the antenna joint to the proper temperature is to have an assistant with a second soldering gun at hand. The application of the second gun to the joint for a few moments will quickly bring the connection up to soldering temperature.

What About A Mast?

If you don't have a sturdy tree, flagpole, barn or house nearby to which to attach your antenna, you will have to put up a wood or metal mast. Don't lose heart--you can build a rugged, inexpensive mast without much trouble which can be put up in 15 minutes by two or three people. (Note: *Never* attach your antenna to a telephone or electric utility pole because of the danger of electrocution from a fallen wire).

There are two types of masts to consider: (1)- *Telescoping TV Mast* - while intended to hold a TV antenna at heights of 20 to 50 feet, these masts can be used to hold up a ham wire antenna providing they are properly guyed. This style of mast comes in 10 foot sections of 18 and 20 gauge wall tubing and is hot-dip galvanized. The mast must be guyed well, at least every 20 feet and better every 10 feet, because without adequate guys the mast has the strength of a piece of wet spaghetti. If a top guy breaks, the mast crashes down in a twisted mess which looks like modern sculpture! (2)- *Wood Mast of 2x3 Spruce*- a more rugged and satisfactory mast can be made of three 20 foot lengths, and one 10 foot length of 2" x 3" lumber. The best kind of wood to use is straight Sitka spruce without knots or splits; this is not always easy to obtain but it is light and strong whereas other woods are heavier and less strong. The next best is straight-grained Douglas fir, or hemlock. All the sections should be carefully selected for as few knots as possible, and be straight (many 20 foot lengths are badly warped).

Building the Mast

Construction details are shown in Figure 5. First give all sections a primer and two finishing coats of first quality outside house paint (follow the instructions on the can). Many hams select a dark green, or

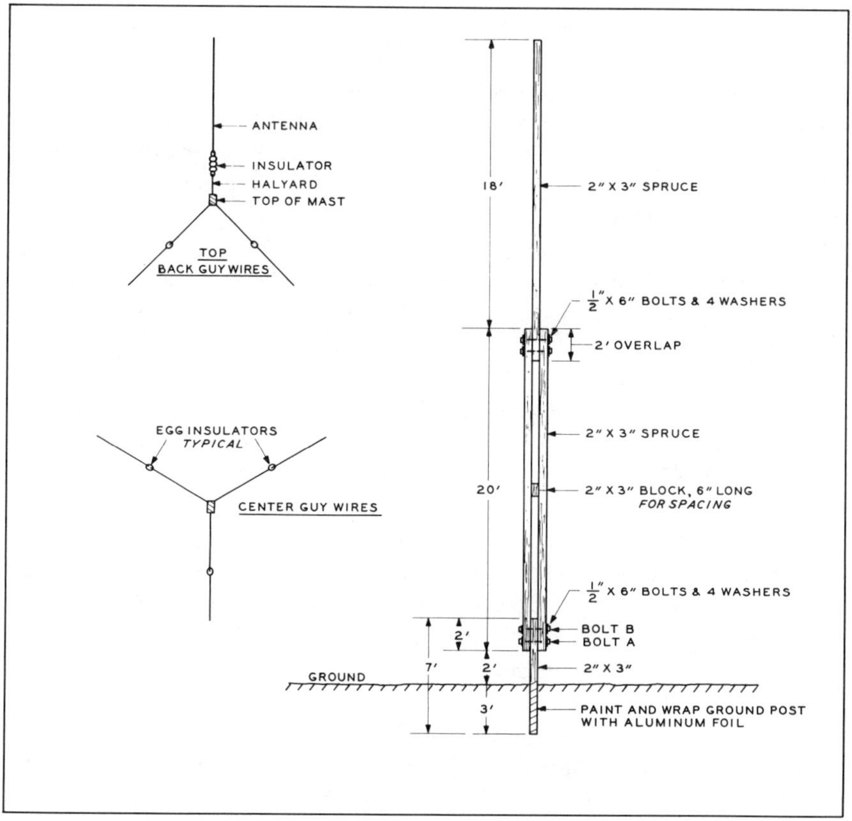

Fig. 5 EASY-TO-BUILD WOOD MAST holds your sky-wire 42 feet in the air. Mast is made up of two 20 foot sections of 2 x 3 spruce. Bottom part of mast is bolted to ground post sunk in the soil. Post is wrapped with aluminum foil to protect it from ground water. Two sets of guy wires insure that mast is stable and will withstand heavy winds.

brown, paint color so the mast blends in with the background and is less conspicuous.

Cut 3 feet off the 10 foot section to make the 7 foot ground post. Or, you can use the entire 10 foot section if you want your mast 43 feet high. (This additional height makes the mast a bit more difficult to erect because of the extra weight). After painting the ground post, wrap the bottom 3½ feet of it with 3 to 4 layers of heavy duty aluminum oven wrap, using vinyl tape to seal each wrapping. This is a *great* new way to keep ground water and moisture from eventually rotting the length in the ground, and is vastly superior to creosote and other alleged rot-preventatives.

Fasten a high quality pulley at the top of your mast by wrapping 3 or 4 turns of strong wire through the pulley fitting and around the mast. Put in two long, thin wood screws below the wire to prevent it from slipping down your mast. Don't drill a hole in the mast to anchor the pulley and don't use an eye-bolt: these schemes weaken the mast and can lead to it breaking at the top. Guy wires should also be wrapped, not fastened through holes drilled in the mast.

Put the rope halyard through the pulley at this point and tie the ends of the rope together to prevent losing the halyard up the mast. Losing a halyard is a tragedy of the first magnitude.(Yes, it has happened to Unlucky Pierre!)

After the sections have been painted,fasten them together as shown in Figure 5. Drill the holes for the 6'' x ½'' galvanized bolts and bolt together all the sections, using washers under the nuts. Nail the small 2'' x 3'' block in the middle of the center section to hold spacing constant.

Take up tightly on the two top bolts since this portion is in final form. Remove the two bottom bolts after making sure they slide in and out easily. Place the aluminum-wrapped bottom of the ground post in the hole you have dug. Fill in and tamp the earth around the post, using a level to keep it vertical (a tilted bottom section means a tilted mast).

Secure the top guy wires by wrapping the wires around the top of the mast (don't drill any holes!) and over the screws which prevent the wires from sliding down. Stranded aluminum wire is recommended for guys. Next, fasten the three center guys where the top section joins the middle section, wrapping the wires around the mast.

Raising the Mast!

Now comes the event you have been waiting for, which is sure to produce some anxious moments! It's time to raise your creation! Unless there is a handy building or tree behind the mast from which an assistant can pull on the back guys to help raise it, you will have to "walk up" the mast. First you use arm power, then a tall step ladder to rest the mast on, and finally another piece of 2'' x 3'' wood about 15 feet long with a U-fitting on the end (so it won't slide off the mast) to push the mast upwards when it is too high for anyone to reach. If there are any extra warm bodies around, have them hold the side guys to minimize swaying (wives and girl friends are useful here). Caution all helpers to remain silent. Jawbone increases the weight of the mast beyond belief!

When the mast reaches a 45 degree angle, bending alarmingly and waving all over the sky, you will swear it looks 100 feet high and think, "We'll *never* get this thing up....." You are 100% normal--everyone thinks that at this crucial point!

However, with one tall, strong ham pushing upwards on the 2" x 3" pole, another standing at the ground post and pulling upward--beads of sweat on his honest brow--and ready to ram home the second bolt (B) which holds the mast vertical--victory is at hand!

Suddenly you are astonished--the mast is up! Shove in the retaining bolt, tighten all nuts, secure the guy wires, and you are in business! Attach the halyard to the insulator and haul up your miracle antenna. The first station you call will give you S9 plus 20 db, for sure!

What About Guy Wires?

Guy wires must be used with wood masts and with telescoping TV masts; in fact, it is wise to use rugged guy wires with any but the strongest type of mast as a safety measure. Guys should be broken every 10 feet with guy wire strain insulators (egg insulators) to prevent unwanted electrical resonances in the wires.

To provide proper support for a mast, three guy wires are spaced 120 degrees around the structure at the mid-point. Two back guys are sufficient at the top, spaced about 70 degrees apart and pulling against the taut antenna halyard which acts as the third guy. A 40 foot mast should be guyed at the 20 foot level, and, of course, at the top.

Guys should be made of wire and not of rope. Heavy, stranded aluminum wire, available at most hardware stores, is easy to work with, stretches very little and is inexpensive. Copper-clad steel wire, copperweld and hard-drawn copper wire are satisfactory but very difficult to work with; if you use any of these, be extremely careful in working with it as a free end may suddenly spring loose, snap around and hit you in the face! Careful! Soft-drawn copper wire, on the other hand, stretches badly under tension and should never be used. Galvanized steel wire is OK for perhaps two years but eventually rusts and must be replaced.

Do not use rope of any description for guys. It is not as strong as wire and it stretches and shrinks to excess. Fiberglas line is not satisfactory for guys, either. Be certain your guy wires do not become kinked as this weakens the wire and can result in it breaking.

Use large turnbuckles at the ground end of the guys so that you can easily adjust the pull of the guys to keep your mast plumb. Your guy wires should be pulled taut but not drum-tight. For safety's sake, pass a safety wire through the turnbuckle so it will not work loose in a windstorm! Remember, too, that galvanized turnbuckles eventually rust; oil your turnbuckles several times a year to increase their life.

Halyards should be replaced every two or three years *before* they wear out to forestall the grief of having to lower your mast to place a new halyard through the pulley.

The Halyard You Select Does Matter!

A wire antenna should be raised into position by using rope halyards. The halyards are tied to the insulators at each end of the antenna, run through pulleys on the mast or support, and then down to the ground. This permits you to raise and lower your antenna easily from the ground for adjustments and maintenance, and perhaps to put up a new kind of antenna later on. *Do not* fasten the antenna directly to a tree or house by simply tying a rope or wire to the support so that you cannot raise and lower it from the ground. This may save a few minutes in erection but it produces hours of frustration later!

When erecting your antenna, tie the two free ends of the halyard together so that one end cannot run up and through the pulley, causing all manner of grief!

Manilla rope, or the more expensive *Dacron* line, should be used for halyards; do not use nylon line as it stretches badly and causes your antenna to sag constantly (it is also expensive). It pays to use 3/8-inch line instead of 1/4-inch line, too, because of the added strength, ease of handling and reliability. Be sure your pulleys are large enough to permit the line to run freely through them; rope swells in rain and sleet and if the pulley fit is too tight to start with, you'll experience difficulty in lowering your antenna when the rope is damp. Many is the ham who discovered that the only way to get his wire antenna down after a heavy rain was to pull hard on the feedline in the center of the antenna -- *not* recommended procedure! Finally, don't use cheap galvanized pulleys as they eventually rust and become inoperative. Use good quality marine-type pulleys of brass or bronze.

What About Wire Halyards?

The thought may cross your mind, "What about using wire instead of rope for the halyards? Wire doesn't shrink."

Fig. 7 COUNTERWEIGHT allows rope halyard to shrink in rain without snapping antenna wire. The best counterweight is a laundry bucket filled with rocks. When antenna is tossed about in a heavy wind, the counterweight rides up and down and much of the movement is absorbed by the weight, leaving the skywire relatively calm.

Right, it doesn't, but there are compelling reasons why it is best not to use wire halyards, despite the fact that a few amateurs do use them:

1)- Wire is difficult to work with and hard to pull taut.
2)- Wire can jam between the side of the pulley and the pulley wheel. When this happens at the top of a 40 foot wood mast you have a problem -- to state it in its mildest terms!
3)- Stainless steel wire is expensive; copper stretches, other wire quickly rusts.
4)- Wire halyards may interfere with the proper electrical functioning of your antenna since under certain conditions the halyards can be coupled to the antenna and alter its electrical length. This is bad news!

The answer is clear: do not use wire halyards.

The Wisdom of Using Counterweights

At the ground end, the rope halyard should not be tied around a tree trunk, branch, cleat or other tie-point, despite the fact that many hams do this. (Thousands of ham antennas fall down every year, too!) Instead, the halyard end should be secured to a *counterweight* of some sort which is suspended a few feet off the ground (Figure 7).

The reason for using a counterweight is so that when the rope halyard shrinks in rain, sleet or snow, or when the antenna and its supports are tossed about in high winds, the antenna wire will not be stretched or snapped, or the halyards become drum-tight and perhaps break under the stress. After two days of rain, a 100 foot length of manilla line can shrink as much as 5 feet! As the rope shrinks, the counterweight simply rises further above the ground and the antenna maintains its proper tension. Later, when the rope dries out and lengthens, the counterweight sinks toward the ground, keeping the proper pull on your antenna.

The Best Counterweight

The best and easiest way to make a counterweight is to use an ordinary laundry bucket and fill it with enough rocks to keep your antenna at its proper height. Drill a few holes in the bottom of the bucket to let rain water drain out.

Old-style sash weights from windows are sometimes used as counterweights but they are difficult to fasten to a halyard and it is not easy to take off, or add, weight.

The counterweight is a safety measure if you have one end of your antenna attached to a tall tree, since the tree will often sway wildly in high winds and jerk your antenna all over the sky. With a counterweight at one end of the antenna, much of the movement is absorbed by the weight, leaving your skywire relatively calm and safe.

Remember: a little thought, time and care spent in selecting your halyards, pulleys, antenna supports and counterweight will pay you dividends over many years of operating!

Check Your Antenna With an SWR Meter

The SWR meter is an inexpensive and handy instrument that can help you determine the actual operation of your antenna. When your new antenna is erected, how can you be sure it is resonant and operating properly? If the choice DX station comes back to some other local amateur, is it just a matter of bad luck for you, or is your year-old antenna at fault? When Joe Blow can hear the weak DX signals and you cannot, is it poor band conditions, or has your five year-old antenna finally "given up the ghost?"

Comparing your signal reports with those of nearby amateurs may be fun but it provides little knowledge or reassurance to the serious amateur who wonders if his antenna is working properly.

The SWR meter can tell you a lot about your antenna. If you log the

SWR readings and re-check them every six months or so, you have a running antenna history that will tell you when something goes wrong with your sky-wire. You can make up an SWR curve of the type shown in Figure 5, Chapter 6 in a few minute's time. The SWR curve will provide a quick picture of antenna resonance. A running check of antenna operation can be made over a period of time and any large deviation in the curve means that something has happened to your antenna system, and you had better look for trouble. A broken joint, a defective feedline, a high resistance joint or a short circuit can cause a quick jump in the SWR reading. If you know what the SWR curve is supposed to be, any deviation from it will be apparent when you repeat your check of the original results.

Make Your Own SWR Charts

It is easy to make up a universal SWR chart for any amateur band with the aid of a copying machine. A master chart is drawn up. The horizontal axis is the frequency range of measurement, plotted every 50 kHz. The vertical axis is the SWR reading (reverse). Leave space on the chart to make a small drawing, or write a description, of the antenna and the date of measurement. This is useful data in the coming days when you may wish to rerun a particular test, or check on the operation of the antenna.

The vertical axis is plotted up to an SWR value of 2.5. In general, most amateur antennas exhibit SWR values less than this, with the minimum SWR reading observed at the resonant frequency of the antenna. As the measurement frequency is moved away from the resonant frequency, the SWR indication will rise.

It is convenient to start the tests at the low frequency end of the band in use. Since you have to feed r-f power through the SWR meter to the antenna to make a measurement, it is a good idea to use the minimum amount of power possible and to run the tests at a time of day when the band is not heavily populated, otherwise you can cause unintentional interference to other stations while you are making your measurements.

Assume you are making an SWR chart for a 20 meter antenna. The first step is to tune your transmitter to 14,000 kHz, or slightly higher in frequency to make sure you are inside the band. The SWR meter is placed in the coaxial line to your antenna in the normal manner. By means of carrier insertion or c-w operation apply a few watts of r-f power to the SWR meter -- just enough to obtain a full scale forward reading on the meter when it is adjusted for maximum sensitivity. Throw the meter switch and read the *reverse* indication. Make a point on the graph to indicate the SWR reverse reading for the 14,000 kHz frequency.

Next, retune the transmitter to 14,050 kHz and repeat the test, logging the reverse SWR reading on the chart, as before. Repeat this test every 50 kHz across the band, up to 14,350 kHz, logging each reverse SWR reading at the proper point on the graph. When you have completed the frequency run, draw a smooth line through the SWR points you have marked on the graph, and your SWR curve is complete.

If you have made accurate readings and if your antenna is working properly you should have a smooth curve, very much like the examples on page 68. The minimum point of the curve is the resonant frequency of the antenna system, which should occur near the center of the amateur band. The SWR curve should rise smoothly and gradually on each side of the resonant frequency. For most amateur antennas, the minimum reverse reading at resonance is below 1.5 and the maximum reverse readings at the band edges range around 2.5, or possibly slightly higher. In the case of a short, loaded whip antenna or a Marconi antenna for 160 or 80 meters, the minimum value of SWR is apt to be higher than these values, but the general shape of the resonance curve remains about the same.

If the point of minimum SWR is too low in frequency, it indicates that your antenna is probably too long. On the other hand, if the point of minimum SWR is too high in frequency, it indicates that your antenna is probably too short. As long as the frequency of minimum SWR reverse reading falls near the center of the band, the antenna length is close enough for proper operation.

The most accurate SWR readings are obtained when a balun or other balancing device is used between a balanced antenna and the coaxial line and when the transmission line is brought away at right angles to the antenna wire so that there is no interaction between the field of the antenna and the line.

The SWR meter is a convenient and informative device to monitor antenna operation and you should check your antenna out every six months or so. Learn to use your SWR meter. It helps you to keep your antenna in tip-top operating condition!

The Radial Ground Wire Revisited

Confusion may exist among some amateurs as to the operation of the radial ground wire and the necessity of having a good *radio ground* connection. As one amateur asked, "What's wrong with an 11 foot long ground lead? Why should I use a radial ground wire?"

The reply is that if a *ground lead* running from radio equipment to an external ground is an appreciable fraction of a wavelength long at the operating frequency, it is an ineffective radio ground. The ground lead, regardless of length, *will* serve as an electric ground against shock and as a lightning ground to conduct static discharges to earth, but its usefulness as a radio ground is impared by length if it is more than approximately 0.05 wavelength long. For 20 meters, as an example, an effective ground lead should not be more than 3½ feet long! For 80 meters, because of the longer wavelength, the ground lead can be as long as 13½ feet and still be effective. Long ground leads are ineffective because they offer too much impedance between the equipment and the ground and in fact may be counter-productive, causing r-f feedback and mysterious TVI problems.

Since a short ground lead may be impractical, the best alternative is to use an artificial ground. Many broadcast stations employ an expensive and large buried radial screen for their ground connection. The less-affluent amateur can achieve the same result by using a resonant radial ground wire, such as discussed at length in this Handbook. If the far end of the resonant radial ground wire is left unconnected, the near end will automatically assume a very low potential to ground. When connected to the radio equipment it simulates a low impedance radio ground connection. *However,* for shock and lightning protection, the use of a ground lead to an external ground in addition to the radial ground wire, is mandatory.

Antennas for the 3 New High Frequency Bands

Three new high frequency bands were allocated to the Amateur Service at the World Administrative Radio Conference, Geneva, Switzerland, in December, 1979. In addition, WARC provided new amateur satellite bands, exclusive frequencies in some shared bands, and other amateur band modifications. This progress resulted from years of preparation by the International Amateur Radio Union and the American Radio Relay League, assisted by other amateur societies, hundreds of volunteers and support from our own and other key governments.

The new hf assignments are: 30 meters (10.1 to 10.15 MHz); 17 meters (18.068 to 18.168 MHz); and 12 meters (24.89 to 24.99 MHz).

Antenna Dimensions

The formulas to determine the length of a halfwave wire antenna are:

$$\text{Antenna length (feet)} = \frac{468}{f(\text{MHz})}$$

$$\text{Antenna length (meters)} = \frac{143}{f(\text{MHz})}$$

Thus dipole antenna lengths (halfwave) for the new bands are: 30 meter band--46'3" (14.1 m); 17 meter band--25'10" (7.87 m); 12 meter band-18'9" (5.72 m).

The formulas for a quarter-wave ground plane antenna are:

$$\text{Antenna or radial length (feet)} = \frac{234}{f(\text{MHz})}$$

$$\text{Antenna or radial length (meters)} = \frac{71.5}{f(\text{MHz})}$$

Accordingly, lengths for the new bands are: 30 meter band--23'0" (7.0 m); 17 meter band--12'11" (3.9 m); 12 meter band--9'4" (2.8 m).

Scaling Antenna Dimensions

Simple wire antennas may be scaled to a close approximation from an existing band to a new band. The following multiplication factors apply for the new amateur bands:

20 meter dimensions x 1.40 = Dimensions for new 30 meter band
15 meter dimensions x 1.17 = Dimensions for new 17 meter band
10 meter dimensions x 1.15 = Dimensions for new 12 meter band

These scaling factors apply to half and quarter-wave dimensions and to matching stubs.

For example, a 20 meter dipole is 33'1" long. It is desired to scale it to the 30 meter band. Accordingly, 33'1" equals 33.08' (see Figure 9). Thus 33.08 x 1.40 = 46.31', or about 46'4". This agrees within one inch of the length computed by formula. Remember the new bands are quite narrow in frequency and an antenna cut for the center of the band will function well over the whole band.

LENGTH FEET (METERS)	30 METER BAND (10.1-10.15 MHz)	17 METER BAND (18.068 - 18.168 MHz)	12 METER BAND (24.89 - 24.99 MHz)
FREE SPACE HALF-WAVE	48' 7" (14.82 M)	27' 2" (8.29 M)	19' 9" (6.03 M)
HALF-WAVE WIRE ELEMENT	46' 3" (14.1 M)	25' 10" (7.87 M)	18' 9" (5.72 M)
QUARTER-WAVE WIRE ELEMENT	23' 0" (7.0 M)	12" 11" (3.9 M)	9' 4" (2.8 M)

Fig. 8 ANTENNA LENGTHS for the three new HF bands. Lengths are for center frequencies in each band. Antennas cut for these dimensions will exhibit low SWR over the entire band since the bands are narrow. Free space dimensions are given for calculation of coaxial transformers.

Trap Dipole Antennas for the New Bands

Since the new hf bands have no harmonic relationship to the existing amateur bands, nor to each other, multi-band dipoles of the type shown in Figure 3, page 75, and Figure 4, page 76, are impractical. The trap dipole, shown in Figure 5, page 78, will work for any band combination. As an example, a two-band dipole for 30 and 40 meter operation is desired. Referring to Figure 5, the high frequency 30 meter center section will be about 46'3" long. The switching traps are each resonant at 10.1 MHz (the low frequency edge of the 30 meter band). Each trap is composed of a 25 pF, 3 kV ceramic capacitor and a coil. The coil inductance is 10 uH. A suitable coil consists of 22 turns No. 14 wire, 1-1/2" (3.8 cm) diameter and 2" (5.1 cm) long. The trap is adjusted to resonance before placing it in the antenna. A dip meter will do the job.

Each end section of the antenna is cut 4' (1.22 m) long and the antenna is raised in the air. Minor length adjustments are made to the end sections to place the resonant frequency in the middle of the 40 meter band. This can be done by trimming the tips equally for lowest SWR on the transmission line at 7.15 MHz.

In like fashion, multi-band trap antennas may be designed for other band combinations. Information for determining coil inductance is given in most communication handbooks, such as the "Radio Handbook" (Howard W. Sams Co., 4300 W. 62nd St., Indianapolis, IN 46268, publisher, or the "ARRL Handbook", American Radio Relay League, 225 Main St., Newington, CT 06111, publisher).

The final step is to determine the length of the tip sections. For a starting point, the overall length of a two-band antenna at the lowest band is taken to be 0.85 of the length of a full-size dipole. Once the antenna is assembled, the tip sections are trimmed to frequency.

DECIMAL	INCH	CENTIMETER
.08	1	2.5
.17	2	5.1
.25	3	7.6
.33	4	10.2
.42	5	12.7
.50	6	15.2
.58	7	17.8
.66	8	20.3
.75	9	22.9
.83	10	25.4
.92	11	27.9
1.00	12	30.5

Figure 9

Conversion of decimal foot (left column) into inches (center) and centimeters (right). For example 3.58' equals 3'7". And 7 inches equals 17.8 cm. One inch equals 30.5 cm.

Multi-Band Antennas for the New Bands

The end-fed antenna design discussed in Chapter 8, Figure 2, will work on the new amateur bands. The radial wire, or wires, should cut to a quarter-wavelength for each band using the formulas given earlier in this chapter. Remember that with this type of antenna a good deal of rf will be present in your station and proper grounding plus the use of radials is essential for normal operation.

The Universal Antenna, discussed in Chapter 12, Figure 1, is suitable without change for operation in the new bands. The coil taps, shown in Figure 3, will have to be determined by trial as discussed on page 164.

INDEX

OTHER BOOKS FOR RADIO AMATEURS, CB OPERATORS, SHORTWAVE LISTENERS, STUDENTS, & EXPERIMENTERS

ALL ABOUT CUBICAL QUAD ANTENNAS, by William I. Orr W6SAI and Stuart D. Cowan W2LX; 112 pages, 75 illustrations.

This well-known classic has been updated to include: new Quad designs; new dimension charts for every type of Quad from 6 to 80 meters; additional gain figures; an analysis of Quad vs. Yagi; Mini and Monster Quad designs; Delta, Swiss, and Birdcage Quads; and an improved Tri-Gamma match to feed a triband Quad efficiently with one transmission line. Also covered are feed systems and tuning procedures for maximum gain and minimum SWR. Much of this data has never before been published.

BEAM ANTENNA HANDBOOK, by William I. Orr W6SAI and Stuart D. Cowan W2LX; 271 pages, 205 illustrations.

This popular new edition gives you: correct dimensions for 6, 10, 15, 20, and 40 meter beams; data on triband and compact beams; the truth about beam height; SWR curves for popular beams from 6 to 40 meters; and comparisons of T-match, Gamma match, and direct feed. Describes tests to confirm if your beam is working properly, tells how to save money by building your own beam and balun, and discusses test instruments and how to use them. A "must" for the serious DX'er!

ALL ABOUT VHF AMATEUR RADIO, by William I. Orr W6SAI; 172 pages, 107 illustrations.

Covers VHF propagation and DX, the VHF repeater and how it works for you, VHF moonbounce work, and amateur satellite communication. Discusses vertical and horizontal mobile antennas, tells almost everything about coaxial cables, and describes VHF beam antennas you can build yourself. Covers SWR measurements, VHF interference and how to suppress it, even care of the Ni-cad battery. This is a complete handbook of VHF radio for amateurs.

THE RADIO AMATEUR ANTENNA HANDBOOK, by William I. Orr W6SAI and Stuart D. Cowan W2LX; 191 pages, 147 illustrations.

This clearly written, easy to understand handbook contains a wealth of information about amateur antennas, from beams to baluns, tuners, and towers. The exclusive "Truth Table" gives you the actual dB gain of 10 popular antenna types. Describes how to build multiband vertical and horizontal antennas, Quads, Delta Quads, Mini-Quads, a Monster Quad, DX "slopers", triband beams, and VHF Quagi and log periodic Yagi beams. Dimensions are given for all antennas in English and Metric units. Tells how antenna height and location affect results and describes efficient antennas for areas with poor ground conductivity. Covers radials, coaxial cable loss, "bargain" coax, baluns, SWR meters, wind loading, tower hazards, and the advantages and disadvantages of crank-up tilt-over towers.